COMPASSIONATE BEASTS

ANIMAL RESCUE LEAGUE
STEP DOWN
PULL

COMPASSIONATE BEASTS

The Quest for Animal Rights

LYLE MUNRO

PRAEGER
Westport, Connecticut
London

Library of Congress Cataloging-in-Publication Data

Munro, Lyle.
Compassionate beasts : the quest for animal rights / Lyle Munro.
p. cm.
Includes bibliographical references and index.
ISBN 0–275–96883–9 (alk. paper)
1. Animal welfare—United States. 2. Animal welfare—Great Britain. 3. Animal welfare—Australia. 4. Animal rights—United States. 5. Animal rights—Great Britain. 6. Animal rights—Australia. 7. Animal rights activists—United States. 8. Animal rights activists—Great Britain. 9. Animal rights activists—Australia. I. Title.
HV4764.M86 2001
179'.3—dc21 99–088488

British Library Cataloguing in Publication Data is available.

Library of Congress Catalog Card Number: 99–088488
ISBN: 0–275–96883–9

First published in 2001

Praeger Publishers, 88 Post Road West, Westport, CT 06881
An imprint of Greenwood Publishing Group, Inc.
www.praeger.com

Printed in the United States of America

The paper used in this book complies with the Permanent Paper Standard issued by the National Information Standards Organization (Z39.48–1984).

10 9 8 7 6 5 4 3 2 1

For Dutchy, Chalkie, and Jenny

Contents

Tables

Tables

Acknowledgements

I was often asked during the course of writing this book why I chose to write about the animal protection movement. My antipathy toward cruelty perhaps had its origins in the decision of my parents to send their reluctant offspring to St. Francis of Asissi primary school, which displayed a prominent picture of the animal-friendly saint. Perhaps too, a gene or two were passed on by my great-grandfather Billy English. When as a nineteen-year-old apprentice jockey he saw how some of the other horses were being mercilessly flayed by their riders, English allowed his mount, Arsenal, to cross the line unwhipped to win the Melbourne Cup of 1886. I therefore record my gratitude for these family influences.

Others to whom I am indebted include my colleagues at Monash University, Gippsland, who have taken an interest in my research. I particularly want to thank Harry Ballis, who encouraged the research from the start, was always ready to rescue me from frequent calamities with my PC, and happily, recommended Praeger as the prospective publisher; Pam Reynolds and Sue Burney-Banfield, who helped with the quantitative data analysis; John Schmuttermaier's frequent brain waves on the use of the computer as a research tool were greatly appreciated; Ray Jureidini, Steve Russell, Keith Wilson, and Peter Farago, who were always willing to read and discuss chapter drafts. Librarians Gary Weston, Xandria Hughes, Malcolm Home and Jane Purton kept an eye out for any material that might have been relevant to the book and were most helpful with literature searches and the like. I also record my gratitude to the library staff in the documentary delivery service at Monash University for tracking down some often obscure references. My research has benefited enormously because of the assistance of these people.

I am also grateful to Monash University for the financial assistance that allowed me to take a semester away from teaching responsibilities to carry out the interviews in the United States and England during the middle of 1996. The university has also provided a number of small financial grants that have paid for the services of expert professional assistance. Here I would like to thank Heather Slater, who typed the hundreds of pages of interview manuscripts so diligently and accurately.

I am indebted to Glenys Oogjes from Animals Australia, Adele Douglass from the American Humane Association, and Joan Court from Animal Rights Cambridge, who opened many doors to individuals and organizations in their respective countries. Without their assistance, much of the quantitative data and many of the key interviews might not have seen the light of day. It goes without saying that I am very grateful to all the survey respondents and interviewees for their cooperation in helping me write what I hope is an accurate profile of the animal protection movement. I would like to thank Suzanne Staszak-Silva, Craig Pelz, and Lisa Webber of Praeger and Elisabetta Linton for their professional guidance during the production process.

On a more personal note, I am grateful to my Canberra friends, including Mike Jones, who once remarked to me that tradespeople he had casually met seemed to talk more about their pets than they did about any other family member, a comment which encouraged me to start this book; Maribelle Young for her early and enthusiastic encouragement of this project, and especially Jan Clark, animal lover *par excellence*, who introduced me to the human and nonhuman companions mentioned in the dedication. These companions, Jenny, Dutchy and Chalkie, have provided the love and friendship that have sustained me during the process of writing *Compassionate Beasts*.

Acronyms

AAHR	Australian Association for Humane Research
AHA	American Humane Association
ACIG	Animal Cruelty Investigation Group
AL(NSW)	Animal Liberation New South Wales
AL(Vic)	Animal Liberation Victoria
ANZFAS	Australian and New Zealand Federation of Animal Societies
ARC	Animal Rights Cambridge
ASIS	Animal and Social Issues Survey
AVA	Australian Veterinary Association
AVS	Australian Vegetarian Association
AWI	Animal Welfare Institute
BALE	Brightlingsea Against Live Exports
BUAV	British Union for the Abolition of Vivisection
CADS	Coalition Against Duck Shooting
DDAL	Doris Day Animal League
FARM	Farm Animal Reform Movement
HSUS	Humane Society of the United States

LACS	League Against Cruel Sports
MRAR	Mountain Residents for Animal Rights
MSPCA	Massachusetts Society for the Prevention of Cruelty to Animals
NAVS	National Anti-Vivisection Society
NCDL	National Canine Defence League
PETA	People for the Ethical Treatment of Animals
RSPCA	Royal Society for the Prevention of Cruelty to Animals
SPCA	Society for the Prevention of Cruelty to Animals
WLPA	World League for the Protection of Animals
WSPA	World Society for the Protection of Animals

PART I

INTRODUCTION: COMPASSIONATE BEASTS

Beasts abstract not.

—John Locke

Compassionate Beasts might seem to be an odd title for a book about animal rights, since the term "beasts" belongs to an earlier era and is rarely used today to refer to nonhuman animals. I use the term "compassionate beasts" to describe the animal protectionists—activists, advocates, and supporters who identify with the animal welfare, animal liberation, or animal rights movements—all of which campaign against the exploitation of animals. In its singular form, "the compassionate beast" serves as an appropriate description of the animal protection movement itself.

The end of a century is a good time to publish a book on the animal rights movement, since many of the movement's most important publishing landmarks have coincidentally occurred at century's end. For example, 1792 saw the publication of Thomas Taylor's satirical essay, *A Vindication of the Rights of Beasts*, which was meant to lampoon Mary Wollstonecraft's book that referred to "women" rather than beasts. And then in 1892, Henry Salt published his seminal monograph, *Animals' Rights Considered in Relation to Social Progress*, a work which Peter Singer, in the preface to the 1980 edition, praised for having anticipated all of the main arguments for the rights of animals. A century later in 1992, the historian E. S. Turner described the progress of the animal welfare cause in words that indicated how far the movement had come in just two centuries:

> It is astonishing how many creatures, from whales to hedgehogs, now have their own pressure groups. In the face of traditional mockery, vegetarianism has made extraordinary advances, not least among the young. The cause of animals has disturbed the calm of company boardrooms, sown self-doubt in universities, driven airlines and airports to show respect for their animal freight, rattled the defenders of ritual slaughter, and caused unwonted and unwanted rifts in bodies like the National Trust. (Turner 1992:318)

Peter Singer's *Animal Liberation* was first published in 1975 and reprinted in 1990, a year that began a decade in which a leading British newspaper described the concept of animal rights as one of "the decade's 10 best ideas" (Appleyard 1995:19). For many people, *Animal Liberation*, first published in 1975, marked the beginning of the modern animal rights movement. But as I explain in Chapter 1, the animal protection movement has a much longer history.

ANIMAL ISSUES FROM A TO Z

In this chapter, I report on the increasing interest of social scientists in animal issues, which literally range from anthropology to zoology. The recently formed International Society for Anthrozoology (ISAZ) is unique as an international organisation in that it focuses exclusively on academic, scientific, and scholarly research in the field of animal-human interactions. To my knowledge, the term "anthrozoology" does not appear in any of the main English-language dictionaries and is rarely found in library catalogues or academic subject listings. It is not surprising that the first Anthrozoology Institute for the study of human-animal interactions was established in England, the birthplace of the animal welfare movement. Although only a few individuals would profess anthrozoology as a discipline, there is an increasing number of professionals, from anthropologists to zoologists, who are doing research in this new interdisciplinary field. For several years now, ISAZ has hosted an annual conference for such people. Thus, in the last decade of the twentieth century, historians, sociologists, psychologists, political scientists, and others in the social sciences and humanities are working on the broad theme of animals and society.

At the 1996 ISAZ conference that I attended in England, one of the speakers, the American psychologist Hal Herzog, claimed that interest in the animal rights movement has virtually petered out, a point that has been challenged by others elsewhere (Rowan & Shapiro 1996). Herzog (1995) claims that just as interest in animal protectionism in England declined between the 1920s and 1970s, there is evidence of a similar decline in the public visibility of animal issues in the United States. His evidence is based on the coverage of animal movement issues in the media and popular periodicals, which he says has not matched the "explosive rise" in the publication of articles in the scientific literature between 1965 and 1985. Public interest in the movement, he suggests, has waned.

In their response to Herzog, Rowan and Shapiro use indicators from popular culture to arrive at a different conclusion (1996). These include television programmes and films that feature animal issues, such as *Free Willy, Babe*, and so on; increases in the donations given to animal protection organisations; a small but noticeable rise in vegetarianism, particularly among young people; and most conspicuously, new developments in the academic world, such as the establishment of centres like the Anthrozoology Institute mentioned above. They also refer to a number of similar centres in the United States and several others that are being developed in America and in Europe. In addition, Rowan and Shapiro cite new social-scientific journals on animal issues as further indications that the field of animal studies is no longer confined to the natural sciences. Finally, they observe that, nowadays, tertiary students would find at least some exposure to animal rights issues difficult to avoid. Thus, it is now possible for university students in the humanities and social sciences to take courses on animals and public policy and to pursue higher degrees with theses on animal rights issues. For example, recent doctoral dissertations on animals-in-society themes include those by Tester (1989), Richards (1990), and Silverstein (1996).

The past decade or so has seen the publication of a substantial number of scholarly books and journals devoted to animal rights issues, as documented by Hogan and Retzel (1995). An *Encyclopedia of Animal Rights and Animal Welfare*, edited by Bekoff and Meaney (1998), is the most recent addition to the field. Hogan and Retzel list about 100 significant books suitable for college libraries that have been published between 1989 and 1995. The predominance of philosophical writing, and the disproportionately large number of books dealing with animals in science, comes as no surprise given the salience of these fields in the history and development of animal rights, particularly in the last two centuries as outlined in Chapter 1.

Scholarly work by sociologists, although still meagre compared to other fields, has increased in the 1990s. One of the first full-length books on animal rights was Tester's *Animals and Society: The Humanity of Animal Rights* (1991), which was based on the author's doctoral dissertation in sociology. While the book stands out in the sociological literature as refreshingly different in its approach to animal rights, it seemed to me to have seriously misrepresented the movement as experienced by activists, advocates, and supporters. Tester argues that the animal rights movement is not about animal well-being at all and is only marginally concerned with animals: "More importantly, it is part of a social project to classify and define humanity. Animals are useful for humans to be able to think human" (Tester 1991:48). Tester claims, without any convincing evidence to support the idea, that "animal rights is not about animals, and as a fetish it is arguing that if we construct a selfhood which is divorced from animals, we will become better human beings" (1991:177). In short, in Tester's view, the animal rights movement is an extreme form of anthropocentrism whose members are concerned only with constructing both a more attractive identity and a sense of superiority over lesser humans who eat, wear, and generally use other animals.

While identity construction is part of the motivation for animal rights activism, it is not the main motive, according to animal protectionists themselves. Insider perspectives are missing in Tester's abstract study. For example, he quotes a single activist—who evidently disliked the term "animal lover"—as representing a group of people who do not necessarily like animals yet campaign on their behalf. Had Tester widened his sample, he may have discovered a whole range of views among animal rights supporters. Tester's thesis is seriously flawed because it is too abstract and speculative. It is an example of strict constructionism concerned with classifications, labels, and discourses. It pays little attention to the experience of social movement participants.

Animals and Society encouraged me to take an entirely different approach to the animal rights movement, one that is based on the insider accounts of movement supporters, advocates, and activists. Fifty-three individuals were interviewed in the United States (17), the United Kingdom (14) and Australia (22), the three main sites for animal protectionism in the world. A profile of these individuals is described below. The interviews, which averaged about forty-five minutes each, were taped and later transcribed for analysis. To distinguish these from other quotations in the text, excerpts from the interviews are *italicised.* I used a survey questionnaire, the Animals and Social Issues Survey (ASIS), for about 350 animal welfare supporters in Australia and a similar sample of American animal protectionists for comparative purposes.[1] In addition, I observed demonstrations, strategy meetings, and protests. Also, I collected a variety of secondary material, including videos, media reports, and movement publications and ephemera.

PROFILE OF ANIMAL PROTECTIONISTS: ACTIVISTS, ADVOCATES, AND SUPPORTERS

The interviewees and the respondents to the survey questionnaire were asked to define themselves as either activists, advocates, or supporters. One or two interviewees questioned the distinctions, but most had little difficulty in making the choice. In the ASIS questionnaire as well, few respondents in this much larger sample of about 350 individuals had any qualms about using these labels to describe their level of activism (Munro 1996a). Unlike in the ASIS questionnaire, however, where supporters and advocates greatly outnumbered the activists, the vast majority of interviewees saw themselves as either activists or advocates, with the sample dividing roughly equally between activists (49 per cent) and advocates (47 per cent). Expressed as a ratio, the activist-advocate profile varied in each country as follows: United Kingdom (3 to 1); United States (1 to 2); and Australia (6 to 5). English and American animal protectors tend to favour grassroots activism and organisational advocacy, respectively, while their Australian counterparts are more likely to practice a hybrid form of activism/advocacy. The reasons for these different patterns of animal protection are explained elsewhere in the book.

That grassroots animal activism is more common in the United Kingdom and

Australia than it is in the United States is in part due to the fact that animal protection organisations employed all of the American respondents. In the English group, most of the interviewees only had marginal links to animal organisations. In contrast, the balance between activism and advocacy was more even in the Australian sample. Even so, it would be unwise to make too much of the activist-advocate distinction in this small sample. Many of the respondents had more than a decade of involvement in the movement at different levels of participation, sometimes as activists involved in direct action, other times as spokespersons for organisations, other times as outsiders offering little more than moral support. As will be explained in the substantive chapters that follow, several Australian animal protectionists, such as Peter Singer, can best be described by the hybrid term activist-advocate.

What is important is that those who identified themselves as either activists or advocates in the interviews and survey questionnaire saw themselves as doing work that required substantial amounts of intellectual, practical, and emotional energy. The average number of years of involvement—15 years for all three countries—suggests a high level of commitment to the animal cause. In addition, fully one-third of the survey respondents and interviewees claimed to be financial members of either the RSPCA or the equivalent SPCA in the United States. As the data from the interviews and ASIS make clear, for many of these people, animal protection had become a way of life that offered daily meaning as well as daily bread.

Personal Background of Animal Protectors

Animal protection organisations, rather than the animal protectionists themselves, were selected as representative of the welfarist, liberationist, and rightist streams in the animal movement. The only attempt to balance the respondents' viewpoints was in the selection of the number of males and females. The sixty to forty sex ratio of female to male in the interview sample does not reflect the high proportion (approximately 80 per cent) of females in the animal movement in the three countries. I deliberately tried to avoid a sample that was too heavily female, as this did not reflect the more even distribution of the sexes in most of the large animal protection organizations selected for study, especially in the United States and the United Kingdom. Australian and English animal protectors tended to be older than their American counterparts. These age differences can be explained by the relatively larger number of retired volunteers and part-time workers in Australia and the United Kingdom. In contrast, most of the American interviewees have full-time careers in animal protection organizations. The majority of interviewees (60 per cent) and respondents to the ASIS questionnaire (73 per cent) were married or in a defacto relationship at the time of the interview, thus challenging "the crazed spinster" label that some critics of the movement are wont to ascribe to "animal lovers."

The 53 interviewees in the three countries have a remarkably similar social profile. The majority (58 per cent) are in full-time professional work, have achieved a high level of education (85 per cent hold a post-school qualification) and live in capital cities or towns rather than in rural areas, although the English protectionists are more likely than their American and Australian counterparts to live in small provincial towns.

Those working in animal protection are intellectually well equipped for the task. The respondents were typically university-educated, with more than 80 per cent holding a first or higher degree such as masters (4 people) or doctorates (5 people). At least three were published authors, and several had written papers for various journals, movement newsletters, and the like. Of the respondents with only a high school education, one individual had worked as a journalist and has continued to write extensively about animal rights issues in the popular press and in movement publications. Most of the those who had completed only a high school education tended to be older people whose formative years were in a period when educational expectations were relatively low.

Lifestyle of Animal Protectors

Although the Australian interviewees were less likely than their American and English counterparts to be childless, 58 per cent of the ASIS respondents—exactly the same percentage as those interviewed—said that they had no children. Childlessness may well be a distinguishing characteristic of animal rights/welfare supporters, since nearly 60 per cent of the sample in the three countries was childless at the time of the interview. In any event, the absence of children combined with employment in white-collar work means that most of the respondents would have a considerable amount of the discretionary time that social movement theorists claim is essential to successful activism (McCarthy & Zald 1973).

The high rate of pet ownership is also a characteristic of animal welfare supporters. Close to 80 per cent of the United Kingdom and United States samples said that they kept companion animals of one sort or the other. The Australian interviewees were not asked about pet ownership, although one can be reasonably certain that the percentage is comparable with the 81 per cent of ASIS respondents who said that they kept companion animals. It is tempting to conclude on the basis of these figures—as some critics of the movement do—that animal protectionists prefer pets to children. A more accurate observation is that companion animals continue to be important in providing "the psychological foundation for the view that some animals at least [are] entitled to moral consideration" (Thomas 1983:119).

Another lifestyle choice that marks people in the animal movement from the rest of the population is their dietary regime. Only 13 per cent of the interviewees said that they were meat eaters, while 28 per cent claimed to be vegan and the remaining 58 per cent vegetarian, which represents a very high rate of meat avoidance even for animal rights adherents. For instance, in the ASIS questionnaire, there were three times as many meat eaters and comparatively few vegans (11 per cent) and vegetarians (51 per cent). However, many of the interviewees worked for a living in animal protection organisations and are probably more committed than those whose daily work is outside the movement. Nonetheless, these figures are a striking contrast to the rest of the population for whom the consumption of meat is the norm. For many people involved in animal protection, diet is the most important lifestyle issue on which individuals can take a stand in their everyday lives. As one would expect, activists were more likely to be vegans, advocates tended to be vegetarians, while supporters in the ASIS questionnaire and the lone supporter among the interviewees were typically meat eaters or semi-vegetarians. The issue of vegetarianism is discussed fully in Chapter 6.

In summary, a composite profile of the typical animal protectionist in these three countries would be white, tertiary-educated urbanites, aged between 30 and 45, half of whom are married or in a defacto relationship, and half of whom are mainly single or never married. The majority are female in paid, professional employment, typically full-time, who are unlike their compatriots in that they have more academic credentials, more pets, fewer children, and are usually vegetarian or vegan.

ORGANISATION OF THE BOOK

Chapters 1–3 in Part I cover the history of the animal movement and the conceptual framework used in the study. Throughout the book, the term "animal movement" will be used as an umbrella term for the more common terms—animal protection, animal liberation movement, animal welfare movement, and animal rights movement. The chapters discuss some of the philosophical and sociological controversies that the movement has generated. Chapter 1 explains the origins of the movement in the United Kingdom, the United States, and Australia as well as explaining the rise of the antivivisection movement in Victorian England. Chapter 2 outlines the campaign orientations in the three case study countries and the tactics and politics of animal protection. This chapter describes the social and organizational basis for the movement's support. While I have tried to maintain an appropriate level of academic detachment in reporting these conflicts, this has not always been successful. In Chapter 3, for instance, I have found it necessary to be as forthright as some of the critics of animal liberation are in discussing some of the controversies that stir people on both sides of the debate. It will be apparent that my own view as to how we ought to treat other animals is on the side of the animal liberationists rather than with those who defend the exploitation of animals. Yet the book is *about* the animal movement, not an argument *for* it; nor has it been my purpose to give "both sides" of the argument on the controversial topics of animal experimentation, intensive farming, hunting, and so on. The book is, after all, a study of the animal rights movement from the perspective of its supporters, advocates, and activists, although from time to time, some of the views of the movement's critics are discussed. I hope then that my position will be read as somewhere between Stephen Clark, the self-confessed zoophile, and the sceptic Keith Tester. Perhaps the designation "critical friend of the movement" best describes the balance that I seek to achieve in the book.

Part II is concerned with grassroots activism, as opposed to organisational advocacy, and includes an analysis of some of the most important campaigns in Australia, the United States and the United Kingdom. This section emphasises the perspectives of movement insiders and the role of women in the movement. Chapter 4 introduces the idea of the activist-advocate as the prototype Australian animal protectionist; how this works institutionally is explained in Chapter 5. The process of becoming an animal activist, and the various forms activism takes, is the focus of Chapter 6. Part II concludes with a discussion in Chapter 7 of the various media that grassroots activists in the case study countries use in their campaigns on behalf of farm animals.

Part III deals with the organisational wing of the animal protection movement in the three case study countries. In Chapter 8, I profile some of the national leaders of the movement in the case study countries and discuss some of the initiatives of multi-issue advocacy organisations that they represent. Chapter 9 deals with one of the most heated issues in the animal rights debate, animal experimentation, the history of which is outlined in Part I. In Chapter 10, I look at campaigns associated with companion and captive animals in the West and in some less-developed countries. These animals provide the basis for building both moral capital and moral outrage as well as bringing in money and resources to animal protection social movement organisations. Chapter 11 covers the issue of recreational hunting, which is analysed for comparative purposes in the three case study countries. Prominent animal advocates-activists are profiled in these chapters to reveal the thinking of the movement's leadership on a range of strategic and tactical issues. In Part IV, I conclude with a review of some of the most effective campaign strategies available to the animal movement in the context of some of the challenges that are likely to confront the movement in the next century.

NOTE

1. The Animal and Social Issues Survey (ASIS) consisted of some nine single pages totalling eighty questions on the themes of attitudes toward animals, the respondents' involvement in animal issues, their views on campaign tactics, and their personal backgrounds, lifestyles, and beliefs on animal and social issues. The response rate was 87 per cent (n=347), well above the usual rate of around 30 per cent for mail-out surveys (Fowler, 1988) and even better than the 84 per cent response rate (n=853) achieved in the similar survey in America by Richards (1992), with which some of the ASIS results were compared.

Chapter 1

A Short History of Animal Protection

> It was no battle of words in which we were engaged but one of ethical conduct.
>
> —Henry Salt in 1921

THE GENEALOGY OF ANIMAL RIGHTS

It is often said that the modern animal rights movement began around 1975, the year that Singer's *Animal Liberation: A New Ethics for our Treatment of Animals* was published. While this is undoubtedly the case, the animal protection movement has much deeper roots that go back to Pythagoras's arguments for vegetarianism in the sixth century B.C. According to Magel (1989), the first publication in English to use the term "rights" in connection with animals was written in 1683 by Thomas Tyron, a Christian theologian. This chapter outlines some of that history by noting the continuity between the "old" humanitarian/welfare movements and the "new" animal liberation/rights movement associated with Singer's name. For more comprehensive historical accounts of our relationship with other animals, there are no better sources than the excellent studies by Thomas (1983) and Ritvo (1987).

Anglo-American Pioneers of Animal Protection

The history of animal protection in the Anglo-American world reveals the continuity of the early campaigns against cruelty with contemporary protests on behalf of animals. It also reveals similarities in the emergence of the movement in the United Kingdom and the United States. The first cruelty law in Anglo-American jurisprudence was inscribed in the statutes of the Puritan founding fathers in 1641. However, as Tannenbaum (1995) points out, the statutes never became law in any of the American colonies, nor even in Massachusetts after it achieved statehood. England first secured the protection of animals in the statute books early in the nineteenth century, although for the rest of the century, animal protection developed along similar lines on both sides of the Atlantic. It was a different story, however, in the eighteenth century, as told by the cultural historian Ann Fairfax Withington (1991) in her book that covers the American Revolution in the period from 1764 to 1776. Cockfighting was a popular pastime in eighteenth-century England, and it was part of colonial life south of New England before Congress banned the practice. But in the political climate of revolutionary America, the vices associated with cockfighting and horse racing posed a threat to the values that colonists needed to resist, or at least morally defeat, English tyranny. "Both activities seemed frivolous and extravagant. They diverted people from work; they produced nothing of benefit to society; and, by encouraging competition and gambling, they weakened community cohesiveness" (Withington 1991:199). Animals were treated better in England than they were in the American colonies, an observation that visitors to both countries often made. The treatment of animals on both sides of the Atlantic came to symbolise the moral virtues, or lack of them, of the protagonists in prerevolutionary America.

The first year of the revolution, 1776, saw Humphry Primatt publish his dissertation on *The Duty of Mercy and the Sin of Cruelty to Brute Animals*, which Turner (1980:11) praised for containing virtually all the intellectual arguments that for more than a century and a half shaped Anglo-American attitudes toward animals. In an introduction to a 1992 edition of the book, Richard Ryder goes even further by noting that Primatt's ideas were remarkably similar to those of leading Anglo-American scholars during the next two hundred years. Primatt even drew a comparison between our treatment of brutes and other races, thus anticipating the link between speciesism and racism that Singer was to make two centuries later:

> Pain is pain, whether it be inflicted on man or on beast; and the creature that suffers it, whether man or beast, being sensible of the misery of it whilst it lasts, suffers *evil*; and the sufferance of evil, unmeritedly, unprovokedly, where no offence has been given, and no good can possibly be answered by it, but merely to exhibit power or gratify malice, is cruelty and injustice in him that occasions it. (Primatt 1992:21)

This 1776 statement against "the sin of cruelty" contains the essential elements of pain, sentience, injustice, and cruelty that became the core motifs of the modern animal protection movement. The influence of Primatt's writings, however, was confined to like-minded souls of his own generation, "a small number of enthusiasts for whom concern for animals had a strong emotional force" (Turner 1980:14). Yet Turner notes that by the end of the eighteenth century, most of "the middling and upper classes" at least believed that morality demanded kindness to the brute creation. It would take another century or more for these changing attitudes to trickle down to the bulk of the population in the Anglo-American world. Much of that task would fall to the genteel patrons of the Society for the Prevention of Cruelty to Animals.

Societies for the Prevention of Cruelty to Animals in the Nineteenth Century

Inspired by Primatt's sentiments, the Royal Society for the Prevention of Cruelty to Animals (RSPCA) found a receptive audience in the Victorian era. Primatt's dissertation emphasised the rational over the emotional and focussed explicitly on animal pain and suffering. England produced the first enduring anticruelty law in 1822 when the Martin Act was passed to prevent cruelty to cattle; in 1835, it was extended to include all domestic animals. The Society for the Prevention of Cruelty to Animals (SPCA) was founded in 1824 in London as the first national animal protection society in the world, although the Liverpool Society for Preventing Wanton Cruelty to Brute Animals, which was set up in 1809, has the distinction of being the world's oldest animal welfare group. In 1840, the national group received royal patronage and became the Royal Society for the Prevention of Cruelty to Animals.

According to Harrison scholars have neglected the history of the RSPCA even though it was one of the most influential "cause groups" in the nineteenth century (1973). Harrison attributes its success to its respectability and to its strategy of never running too far ahead of public opinion, especially among the more respectable members of the public. The RSPCA derived its philosophy, rhetoric, and moral force from evangelicalism; the objects of its campaigns against cruelty were almost exclusively the cruel sports of the lower orders. In Thompson's view, the organisation was swimming with the tide, not creating it, as it was "just one weapon in a whole armoury which included Sunday schools, temperance, popular education, and personal hygiene, that was intended to persuade or compel the working classes to adopt approved standards of behaviour rooted in religious belief" (Thompson 1988:280).

Another historian dates the humanitarian tide as having begun in the previous century: "Throughout the eighteenth century, humanitarian reform had played a major causal role in [the] cultural reconstruction of pain, identifying a range of formerly unquestioned practices as unacceptable cruelties and demanding that virtuous people, men and women of sensibility, endeavour to put a stop to such practices" (Haltunen 1995:318). These included blood sports, public executions, treatment of the insane, punishment by flogging in the armed services, corporal punishment of children, and sport that caused serious injury. Still another historian noted that eighteenth-century women empathised more with animals than they did with peasants or slaves. Barker-Benfield sees the eighteenth century as an age of sensibility in which women campaigned against male barbarity: "From Margaret Cavendish through Francis Power Cobbe, women made the connection between men's treatment of animals and their treatment of women" (Barker-Benfield 1992:232). Cobbe wrote an article in 1878 called "Wife-Torture in England" in which she put men's abuse of women on par with vivisection. She condemned proponents of vivisection for experimenting on the nerves of living dogs and torturing dozens of cats in a single experiment.

Animal protectors in the RSPCA had to work hard to change the way that "the lower orders" treated animals. In the early 1800s, legislation against bull-baiting did not invoke the cruelty argument because the concept would not resonate in a predominantly rural England, which depended on the daily slaughter of animals (Harrison 1973:786). For example, some thought the House of Commons reached a low ebb in 1880 when it had a bill before it to abolish bull-baiting (Ritvo 1987:125). By the end of the century, argues Harrison, kindness to animals was the widespread expectation of an urban society that had become used to the RSPCA's strong appeals for compassion towards "the brute creation." One humanitarian crusader, W. J. Stillman noted approvingly how far animal protection had come over the past one hundred years; in Anglo-Saxon countries especially, tenderness towards sentient creatures, he observed, was now commonplace (Ritvo 1987:126). And as Ryder (1989) points out, the RSPCA in the second half of the nineteenth century succeeded in promoting a number of important legal protections for animals, such as protecting dogs used to draw carts, wild birds, laboratory animals, animals in transit, and even wild animals.

Not withstanding these achievements, and similar ones in the United States, the organisation on both sides of the Atlantic was, as a creature of the Victorian age, characterised by a belief in "kindness, mildness and self-restraint," virtues unlikely to upset the status quo (Turner 1980:69). It became glaringly obvious that on some issues, most notably the vexed question of hunting with hounds, the RSPCA would not challenge the proclivities of the upper classes. Working-class habits were a different matter as the records of the society's inaugural meeting in 1824 revealed. Ritvo notes that the chairman made it clear to members that the society's purpose was not only "to prevent the exercise of cruelty towards animals, but to spread amongst the lower orders of the people . . . a degree of moral feeling which would compel them to think and act like those of a superior class" (1987:135).

According to Turner, for most of the nineteenth century the SPCAs in North America followed in the footsteps of the English RSPCA by cultivating a conservative, middle-class constituency and by keeping women from ascending to leadership positions in the organisation: "The RSPCA largely expelled the unconventional; its American counterparts simply ignored them" (Turner 1980: 53). The American Society for the Prevention of Cruelty to Animals (ASPCA) was established in 1866. In the next three years, SPCAs spread to Boston, Philadelphia, and San Fransisco. The founder of the ASPCA, Henry Bergh, persuaded the state legislature of New York to include all animals in the 1867 animal protection statute. The first section of the statute has been incorporated into many American state anticruelty laws, and Tannenbaum notes that even the wording closely resembles the language of present-day statutes:

> If any person shall over-drive, over-load, torture, torment, deprive of necessary sustenance, or unnecessarily or cruelly beat, or needlessly mutilate or kill, or cause or procure to be over-driven, over-loaded, tortured . . . any living creature, every such offender shall, for every such offense, be guilty of a misdemeanour." (Tannenbaum 1995:567)

Although legislation to protect animals was first enacted in England, animal protection could be described as an Anglo-American tradition. According to Turner, the Anglo-American world in the nineteenth century was a separate cultural entity within the larger European civilization. Thus, it is not surprising that animal protection in both countries followed a similar pattern. Worster also writes of a distinctive Anglo-American tradition in the ecology movement in the late twentieth century, which while "never wholly a consensus, but withal a single dialogue carried on in a single tongue" (1977:ix).

Australia's early efforts in animal protection were also part of this dialogue. An Australian RSPCA was established in 1891. By the end of the nineteenth century, each of the colonies had its own society modelled after the English parent organization. Like its Anglo-American counterparts, the Australian RSPCA consisted of predominantly middle-class urbanites, although in the Australian case, the RSPCA attracted affluent people from rural areas as well. Yet aside from these similarities, the animal protection movement in Australia developed along different lines and in ways that were unique.

Cruelty in the Australian Bush

The most comprehensive study of the early history of the animal protection movement in Australia is Jennifer MacCulloch's (1993) analysis of the period 1880–1930, in which she documents the activities of the Animal Protection Society of New South Wales (APS), founded in 1873, and the Women's Society for the Prevention of Cruelty to Animals, established in 1886. The APS's moral purpose was not unlike the RSPCA's in that it derived its motivation from "the prevalence of cruelty to animals amongst a certain class of persons in this community [which] has excited the indignation and disgust of the educated and humane" (MacCulloch 1993:43). Opposition to cruelty was to be the driving force of the early animal protectionists in their campaigns to save both domestic and wild animals in Australia. While the brutality of working-class men, particularly toward street horses, had been the target of the early campaigners, most of whom were women, the antiplumage campaign targetted women for their indirect cruelty to birds. The antiplumage campaign became a synonym for anitcruelty and united both animal protectionists and preservationists. One of

MacCulloch's most important findings was that these two movements were irrevocably intertwined and, in the popular imagination, synonomous.

It is perhaps not surprising in a country with a relatively small population that animal protectionists and preservationists would find it necessary to join forces in their respective causes. But according to MacCulloch, the preservationists were keenly conscious of the disadvantages of being associated with "the emotionalism, hysteria and religiosity of the animal protection movement" (1993:202). More importantly, then as now, both movements had different purposes and ideologies; the preservationists focussed their campaigns on saving species and habitats, and they thought little of the animal protectionists' concern for preventing cruelty to individual animals. MacCulloch argues, however, that the animal protection movement's lasting legacy was shaping the means by which communication about nature with the public was possible. According to MacCulloch, the preservationists realized that however much they were out of kilter with the animal protectionists, the concept of cruelty was needed to mobilise a sympathetic, sensitive public into action. The early alliance of the two movements in Australia was therefore a marriage of convenience since the preservationists needed the moral appeal of their partner's anticruelty focus. In her conclusion, MacCulloch describes what still rings true a century later: "For no matter how affecting, or even tragic, it was to witness the destruction of a single, beautiful tree or scenic area, it lacked the pathos of cruelty to animals. A tree was a living thing but it did not bleed, it did not suffer, it did not have babies" (1993:369).

Animal protectionists have always known the power of cruelty as a moral weapon, yet it was rarely widely invoked on behalf of native Australian fauna. It featured prominently in the antiplumage campaign and most effectively in what was to become the preservation movement's symbol of success, the campaign to protect the koala. Of all Australian fauna, the koala provided the highest-quality fur. Hunting them for profit led to the destruction of vast numbers of the animals. Although numbers are imprecise, MacCulloch reports that one leading fur company held over one million koala skins for export to Russia in 1896.

Animal protectionists and preservationists had little difficulty promoting the virtues of the "bushland baby and toy" to city people as "a morally blameless infant which lived on gum leaves and harmed no-one and nothing" (MacCulloch 1993:299). To many rural people, however, the koalas were vermin to be exterminated. To counter this hostility, the campaigners were obliged to transform the koala's status from commercial success to national pet. The modern meaning of koala, is as an emotional rather than commercial resource, much the same as the social value of children changed from an economic to an emotional resource after the Industrial Revolution (Zelizer 1985). But in the early part of the century, koalas were hunted as pests. If people were to see them as anything other than vermin, an effective campaign would need to be waged. Cruelty was the answer, for how could anyone "murder" this quintessentially cute and cuddly creature?

The koala was therefore constructed as a lovable pet, an Australian teddy bear, which one reporter described as a "charming little creature, a more perfect natural pet is not to be found anywhere in the world" (MacCulloch 1993:300). MacCulloch points out that this construction was only possible once the koala had ceased to be of commercial value, after overhunting had decimated their numbers. For this reason, she argues, the koala was a good choice as a symbol for the preservation movement. The koala was "an amalgam of the values and messages of the animal protection and preservation movements . . . the conduit which connected the two in the popular mind" (MacCulloch 1993:301). According to MacCulloch, animal protectionists needed the preservationists since the animal lobby had virtually run out of steam by 1914. MacCulloch concludes that this was due to the movement's transformation from a social-reform lobby to a group of pet enthusiasts who emphasised the genteel promotion of kindness to domestic animals, especially cats and dogs: "This loss of purpose both mirrored and was reinforced by the growing feminisation of the cause. Increasingly, the cause of animal protection was given over to women, and subsequently, children" (MacCulloch 1993:45–6).

Elsewhere within the animal protection movement itself, unity of purpose has not always been easily achieved. At least in the nineteenth century, the RSPCA provided a solid foundation for a united voice to be raised in defence of animals. What there was of a single voice in the movement, however, was to dissipate in the first half of the twentieth century. Part of the reason was the growing tension within the animal protection movement between moderate welfarists in the RSPCA and SPCAs and the abolitionists in the antivivisection movement. According to French, by 1876, the RSPCA had distanced itself from "the hysteria and sensationalism" in the antivivisection movement and even took to discrediting it publicly (1975:82). As the tension between the two movements remains unresolved at the end of the twentieth century, it is worth outlining the history of antivivisectionism.

THE EARLY OPPOSITION TO ANIMAL EXPERIMENTATION

> Time will disperse with certain hand and sure
> This swarm of buzzing idlers, whose obscure
> Humanitarians foul humanity. (Zoophilist, 1835, cited in Elston 1990:259)

These inflammatory words were among the many directed against those who campaigned against the use of live animals in scientific experiments in America and England in the nineteenth century. The term vivisection originally meant the dissection of, or the surgical intervention upon, living animals for research purposes (Maehle & Troehler 1990:14). Vivisection has always been a controversial scientific procedure, but it was not until the second half of the seventeenth century that it became a moral issue.

In the first half of the seventeenth century, Descartes maintained that animals were *automata* devoid of rationality, soul, and feeling. Unlike many of his disciples, Descartes did not believe that animals were entirely devoid of sentience, and toward the end of his life he conceded the possibility of some inferior feelings in brutes. But the early vivisectionists who followed the Cartesian concept of the "beast-machine" believed that they could not harm animals; the cries of animals subjected to vivisection were interpreted as no different than the sounds that an object emitted when struck by a hammer. Here was a rock-solid rationalization for exempting humans from any qualms that they might have about their treatment of animals. But by the end of the seventeenth century, the rock had become increasingly shaky as educated individuals, vivisectors among them, insisted that animal experimentation was cruel.

Robert Hooke (1635–1703) for example, experimented on a dog, during which he cut away the thorax and diaphragm of the living creature to observe the motions of the animal's heart and lungs. He managed to keep the dog alive for more than an hour by pumping air into its windpipe by means of a pair of bellows. His observation of the animal during the process moved Hooke to vow never to repeat the experiment "because of the torture of the creature" (Maehle & Troehler 1990:23). While there were scientists who found vivisection abhorrent, the majority of vivisectors rarely mentioned the torture they inflicted on their living tools.

The Vivisection Debate

Samuel Johnson (1709–1784) rejected Cartesianism and believed, no doubt by observing his own pets, that animals felt pain and had feelings. His attack in 1758 on physicians who experimented on animals rates as the fiercest denunciation up to that time: "Among the inferior Professors of medical knowledge, is a race of wretches, whose lives are only varied by varieties of cruelty" (Maehle & Troehler 1990:32). His arguments against vivisection, while not original, were effective because they were anthropocentric. First, he claimed that cruelty toward animals may lead to cruelty toward men, and second, he denied any therapeutic consequences of animal experimentation, suggesting that "these horrid operations . . . make the physician more dreadful than the gout or stone" (Maehle & Troehler 1990:33).

William Hogarth's depiction of the fate of Tom Nero in *The Four Stages of Cruelty* (1750) graphically captured these sentiments in pictorial form. Nero progresses from torturing animals in the streets of London to beating a horse to killing his mistress, for which he is hanged. In the fourth stage of the tale, the still breathing body of Nero is the subject of an anatomy lecture, during which the surgeon's mastiff consumes his entrails. Immanuel Kant, who argued that the considerate treatment of animals was a duty to man himself, apparently used Hogarth's prints to illustrate in his own lectures the widely held idea that cruelty to animals may lead to cruelty against one's fellows.

But it was left to Jeremy Bentham to produce what was probably the first serious counterargument to Kant's doctrine that "our duties towards animals are merely indirect duties towards humanity" (Kant 1930:230). Kant evidently believed that painful experiments on animals were to be avoided if the results could be otherwise obtained; but it was what the experiments did to the experimenter rather than to the animal that chiefly concerned Kant. In stark contrast to Kant's human-centred, contractual doctrine, Bentham's (1789) oft-quoted dictum—"the question is not, Can they *reason*? nor, Can they *talk*? but, Can they *suffer*?"—called for an entirely different relationship between humans and animals. This right of animals to be treated humanely as sentient beings was to be the basis for Singer's (1975) *Animal Liberation* almost two centuries later.

Victorian Values and Vivisection

It was not until the 1870s that the vivisection controversy reached its peak in Victorian England: "What elevated antivivisectionism from a mostly latent sense of outrage into a ferocious public agitation was the large-scale importation of experimental physiology into Britain and the United States after 1870" (Turner 1980:89).

Articles, letters, advertisements, books, essay contests, placards, flyers, pamphlets, periodicals, novels, and paintings were among the many ways the moral atrocities of animal experimentation—as described openly in Sanderson's *Handbook for the Physiological Laboratory* of 1873—were widely publicised. As French points out, the printed word was the single most important medium for mobilising antivivisection support (1975:254).

Opposition to vivisection was such that the biomedical establishment was compelled to organize its members against the twin charges that animal experimentation was both cruel and useless. Rupke points out that the antivivisectionists suffered a setback in 1881 when the scientists rallied in London for the International Medical Congress and unanimously endorsed animal experimentation as indispensable to medicine and therefore "in the interests of man and of animals" (1990:191–2). Insult was added to injury when the journal *Nature* criticized leading antivivisectionists for their apparent hypocrisy. Francis Power Cobbe, for example, was ridiculed for wearing ostrich feathers ("plucked from the *living bird*") and for being seen in public carrying an ivory-handled umbrella ("*cut out* of the dying elephant's jaw"). The people-in-glass-houses taunt would be used whenever animal protectionists could be faulted for not living up to the high moral standards of their most militant colleagues.

Rupke suggests that the defence of vivisection by reference to the cruelties of blood sports was no more than an attack on the middle-class and aristocratic membership of the antivivisectionist societies (1990:203). However, the single most important motive for defending animal experiments, he argues, was that the antivivisectionists threatened not just the progress of biomedical science, but the status of the scientists whose position and privilege had steadily improved during the Victorian period (1990:198).

Nor was opposition to vivisection confined to animal protectionists. The biomedical fraternity also had to contend with the less strident but sometimes more effective arguments of the Humanitarian League (1891–1919), founded by Henry Salt. The league argued against the use of animals in experiments because they were too much like humans. It also opposed the use of a vaccine made from calf lymph because it was an example of "the animalization of humans" (Weinbren 1994:99). According to Weinbren, the league introduced a fresh perspective to the antivivisection debate by inserting new issues, such as the role of state intervention, the importance of education, and the promotion of a wholesome lifestyle (1994:98). The league's approach, while relying more on education than vilification, was nonetheless designed to shame those who they thought were cruel to animals.

The experimentalists used vilification as well. In America, "the nightmare humanitarians of antivivisection" (Fleming 1987:145) were subjected to extreme abuse by the experimentalists, whose rhetoric of vilification has continued well into the twentieth century. The historian Donald Fleming penned one of the most colourful examples in the 1950s in his study of the rise of modern medicine in the United States. To this modern-day defender of animal experimentation, the antivivisection movement of the nineteenth century was

> a world alliance of overlapping animal lovers, vegetarians, anti-Listerians, antibacteriologists, and anti-inoculationists; soft muddle-headed women, cold-blooded Amazons and publicity seekers of both sexes, bluff advocates of the old-fashioned godliness and soap-and-water but not carbolic-acid cleanliness, maudlin religionists not averse to blasphemy, medical sectaries, and academic and professional cast-offs—all engaged in a travesty of the intellectual life, snooping about laboratories and ransacking the pages of scientific journals for ammunition. (Fleming 1954/1987:147)

Turner (1980:93) points out that American antivivisectionists relied almost exclusively on the literature of their English colleagues, who as with animal welfare generally, took the lead in campaigning against cruelty. On one matter, however, the Americans proved to be the more innovative. America's vivisectors saw themselves as rational men of science whose work on behalf of humanity was being retarded by middle-class, city-based female "cranks" in humane societies. In 1906, a prominent American neurologist, Charles Loomis Dana, foresaw "catastrophe" for science if the antivivisectionists succeeded in their plans to regulate laboratories.

His solution was to pronounce the opponents of animal experimentation victims of zoophil-psychosis, an illness he diagnosed as among the "obsessive insanities" (Buettinger 1993:5). Women—some of whom it was alleged coddled their pets and loved them more than babies—were held to be more susceptible than men were to the disease. Buettinger refers to one physician who maintained that zoophil-psychosis affected people without children and without a serious occupation in life. Not surprisingly, women who joined animal protection societies, or worse, campaigned against vivisection, were easily portrayed as afflicted by the malady. The *New York Times* of March 1909 explained: "This is because animals make a small and strictly limited demand for affection and care, and therefore are favorite objects of those with little to bestow" (Buettinger 1993:8) The newspaper's verdict has become a popular refrain among critics of the present-day animal movement.

By 1914, notes Buettinger, the charge of zoophil-psychosis had become commonplace despite the lack of any scientific evidence for the malady. And although antivivisectionists refused to be intimidated by the label or the labellers, the movement began to run out of steam by 1925 (Buettinger 1993:10). Yet it was not Dana's invention of zoophil-psychosis that retarded the progress of the antivivisectionists.

Lederer argues that the causes were more varied, including the antivivisectionists' own excesses, their continued support for obsolete medical therapies, their making of unhelpful alliances, and their strident attacks on medical progress in understanding disease (1990:253). Articles in quality press such as the *New York Times*, like the example quoted above, important as they were, did not have the same impact as an essay that appeared in the Encyclopedia Britannica of 1910–1911. The piece carried the imprimatur of the Britannica, which presented a six-page, 9,000-word argument supporting vivisection by Stephen Paget, honorary secretary of the Research Defence Society. Paget invited readers to compare the fate of the 86,277 research animals in Great Britain in 1909 to the much worse predicament of "an equal number of the same kinds of animals, either in a state of nature, or kept for sport, or used for the service of human profit or amusement" (Cockburn 1996:28). Cockburn suggests that the article's inclusion in the eleventh edition was in response to the strength of the antivivisection movement at that time. It is more likely that the editors were confident of the article being in tune with a public that had already been, for the most part, won over by the case for vivisection.

In any event, according to Lederer, by 1914, moderate animal protectionists sympathised with medical researchers about the excesses of their more militant colleagues. Antivivisectionism had lost much of the moral capital it had enjoyed in the 1870s, but by no means did the movement disappear all together.

ANIMAL PROTECTION IN THE TWENTIETH CENTURY

According to Ryder (1989), royal support of the RSPCA continued until the end of the nineteenth century but declined after World War I. With it went the fortunes of the RSPCA. Elsewhere, Ryder (1979:11) suggests that from 1918 to the 1960s, serious intellectual debate on animals' rights was at a low ebb. And when one looks at the literature on the animal protection movement in the first half of the twentieth century, there is an apparent falling off in interest by scholars.[1]

On the other hand, social movements professing to be against *all* cruelty were much in evidence at the turn of the century. Salt's Humanitaran League is the best example of a multipurpose anticruelty campaign. Its mission included reforming Poor, Criminal, and Prison Law, stopping cruelty to animals in vivisection or in slaughterhouses, and halting the trade in feathers, blood sports, and the "evil trade" that involved the shipping of live cattle abroad (Weinbren 1994: 88). Unlike the RSPCA, which was reluctant to harass the hunting fraternity, the Humanitarian League campaigned vigorously against blood sports. In 1924, it established the League for the Prohibition of Cruel Sports, the most recent campaigns of which as the League Against Cruel Sports are described in Chapter 11. Salt saw his organisation as "a fighting, not a talking Society" (Hendrick & Hendrick 1989:46). He had little patience for the hairsplitting over the meaning of "rights" that still preoccupies philosophers today.

Salt believed, and many modern animal welfare/liberation organisations agree, that animal welfare reform can best be achieved via education and legislation. His firm faith in education was matched by his lack of faith in established religion. His only religion was his self-styled *Creed of Kinship*, which he hoped would have an educational, civilizing effect on those who considered its principles: "a belief that in years to come there will be a recognition of the brotherhood between man and man, nation with nation, human and sub-human, which will transform a state of savagery as we have it, into one of civilization" (Winsten 1951:203).

Legislation, the second weapon in Salt's reformist armoury, was by no means the lesser strategy, for Salt argued that it was more important than preventive cruelty was. Legislation—such as the reforms that abolished slavery and established the Factory Acts—provided "the record, the register, of the moral sense of the community" (Salt 1980:124). Time and again, animal protection organisations that advocate moderate, piecemeal reforms in animal welfare echoed Salt's sentiments. Not surprisingly, the most striking example of an organisation choosing the legislative track as a strategy against cruelty is the League Against Cruel Sports, the group that Salt's Humanitarian League established in 1924.

Thus, it would be mistaken to conclude that the animal protection movement had run entirely out of steam in the first three-quarters of the twentieth century. Organisations like the RSPCA did not disappear, nor was it the case that no new groups emerged during this comparatively quiet period. For example, in one list of major British animal protection organisations that were formed in the nineteenth and twentieth centuries, half were founded between 1900 and 1969 (Garner 1993:43). Furthermore, Lederer (1991, 1992) has described the durability of the antivivisection movement during this period and its capacity to exert pressure on the medical research fraternity. Moreover, as long as there were individuals like Salt who devoted their lives to the humanitarian cause, the animal movement would continue to move, however slowly. Small numbers of supporters would join the true believers as in any social movement whose grievances had not been resolved. Eminent individuals such as Gandhi—who attributed his conversion to vegetarianism to Salt—would add lustre and, more importantly, moral capital to the movement. Gandhi's belief in *Ahisma*—nonviolence, of causing no harm to humans or animals—is the cornerstone of Hindu ethics. It continues to guide animal liberation philosophers, such as Tom Regan, and grassroots activists as far apart geographically as Joan Court in Cambridge and Anne Elliott in the Blue Mountains near Sydney. The commitment of these individuals to nonviolence is described in the interviews in the relevant chapters which follow. But first, there is one episode in the first half of twentieth century animal protection that demands explanation.

Nazi Animal Protection Policies

Nothing contrasts more starkly with the Gandhian approach to animal rights than the politics of animal protection in Nazi Germany. It is surely one of the ironies of history—and an episode of acute discomfort to animal protectionists—that some of the most progressive animal protection legislation this century was enacted in Germany during the reign of the Nazis. Just as animal protectionists are pleased to invoke Gandhi's vegetarianism but not Hitler's, there exists little more than an explanatory footnote to the seemingly incongruous attitude of one of history's most genocidal regimes to the world of nature. However, the recent debate sparked by Arluke and Sax (1992) provides a deeper insight into this apparent contradiction. Arluke and Sax suggest three reasons for the paradox. First, the Nazis' love of animals may have been a psychological response or coping device to counter their inhumanity toward "inferior" humans. Second, their professed love of animals may have been a convenient guise to attack inconvenient people like Jews, homosexuals, and gypsies. And third, the Nazis may have blurred the boundaries between humans and animals so as to be able to designate certain undesirable humans as lower in moral worth than animals were.

Elsewhere, Arluke and Sanders (1996) claim that the Nazis' blurring of the animal/human boundary, and their alteration of the conventional distinctions of what in the West constitutes animal and human was achieved in three ways: the moral elevation of animals, their identification with them as noble beasts, and the animalization of certain humans.

While critics of modern-day animal protectionists are quick to connect their activities with the Nazis' attitudes toward animals, no mainstream animal liberationist would accept the boundary work that the Nazis practised. First, the moral "elevation" of animals does not usually feature in animal liberation campaigns. There is nothing in the contemporary animal movement approaching the Nazis' cult worship of animals or the distinction made in the Third Reich between good (Aryans) and bad animals (Jews and other "inferior races"). Nor do animal liberationists claim, as the Nazis did, that animals like the "sacred" pig were moral beings. Second, Nazi identification with animals at a personal level is not unusual; most of the leading Nazis, from Hitler down, kept pets and professed their mutual devotion in the same way that all pet owners typically do. Hitler and Goebbels were different in that they preferred animals to humans, a preference that animal liberationists rarely expressed. (Of the fifty or so individuals interviewed for this book, only two confessed to feeling this way).

According to Arluke and Sanders, the Nazis identified with animals that symbolised "ideal" qualities such as fearlessness, loyalty, obedience, and strength (1996:165). Politically, however, identifying with animals was important as Nazi propaganda because it directed people's compassion away from suffering Jews to innocent animals that the Nazis claimed were victims of Jewish butchers and vivisectors. And Goering, the avid hunter, promised "to commit to concentration camps those who still think they can continue to treat animals as inanimate property" (Arluke & Sax 1992:7). While Goering's own love of hunting did not constrain his contempt for such people, it symbolized the fundamental contradictions inherent in Nazi attitudes toward animals and humans. Further, the animalization of humans as "vermin," "beasts," and "wild animals" is a technique of demonization not confined to the Nazis. Animal liberationists constantly challenge the unthinking labelling of criminals and deviants as "animals" by pointing out the injustice of attributing to animals what humans alone are guilty of.

According to Luc Ferry, the animal protection law (*Tierschutzgesetz*) of 24 November 1933 and the law for the protection of nature (*Reichsnaturschutzgesetz*) two years later represented the first time in history that an attempt to reconcile political and ecological interests was made (1995:92). Compared to other animal protection laws in Europe, the Nazis' *Tierschutzgesetz* stood out as a rejection of anthropocentrism and as a principle that animals were to be protected in their own right and not as a means to shield humans from the ill effects of cruelty. Arluke and Sax (1992:9) dispute this however. They argue that the Nazi laws did not go much beyond those in Britain, the country that had pioneered animal protection. Phillips and Sechzer (1989:12) support this view when they point to the similarities between the Nazi law of 1934 and the British law of 1876, which they explain as a consequence of Hitler's selective admiration for British values. What made the German law different was the severity of the punishment meted out to those found guilty of mistreating animals (Arluke & Sax 1992; Brunner 1995). Arluke and Sax give the example of one such penalty: a fine plus two years imprisonment for the "rough mistreatment" of a dog.

What continues to intrigue scholars, however, is that the purpose of Nazi animal protection was, in the words of the *Tierschutzgesetz*, "to awaken and strengthen compassion as one of the highest moral values of the German people" (Arluke & Sax 1992:8). Among the first reforms that the Nazis implemented when they came to power in 1933 was the law relating to Jewish ritual slaughter. Laws regulating animal experimentation, protecting wildlife, and forbidding the boiling of crabs, lobsters, and other crustaceans soon followed. The Arluke and Sax paper stimulated several responses to these contradictions—the idea of protecting animals while murdering people, the avoidance of meat but not blood sports, the receptivity of the Nazi regime's evil intentions in a highly cultivated society, and so on. Some commentators have warned of the danger of making too much of the Nazi paradox. Herzog, for example, makes the compelling point that "moral inconsistency is the hallmark of our relations with other species" (1993:82).

The Modern Animal Liberation Movement

As noted at the beginning of this chapter, Singer's *Animal Liberation* (1975) is frequently cited as the catalyst that launched the modern animal rights/liberation movement. Jasper and Nelkin (1992) for instance, suggest that every animal rights activist is familiar with the book they describe as "a bible of the movement."[2] While Singer's many books, articles, and interviews are important, they are, along with the work of other philosophers I discuss in Chapter 3, only part of the intellectual toolkit of the modern animal movement. Philosophers like Singer and Regan, as Jasper and Nelkin remind us, served as midwives to the animal rights movement in the late 1970s by providing an explicit ideology that would link people's strong emotional feelings about animals to action on their behalf (1992). Jasper and Nelkin claim that the animal rights agenda "won the hearts and the minds of people seeking ways to articulate their growing concern for animals" (1992:90). The bulk of *Compassionate Beasts* is concerned with showing how animal protectionists in the United Kingdom, the United States, and Australia are engaged in a quest that Regan (1987) called "the struggle for animal rights." In my terminology, animal welfare advocates in the suites represent the "mind" of the movement while grassroots activists in the streets provide the movement's passion or "heart." While these distinctions are not always clear-cut in the everyday struggles of animal rights activism and advocacy, they do reflect how many animal protectionists describe their involvement in the movement at the end of the twentieth century. It is these insider accounts that the rest of the book seeks to represent.

NOTES

1. For example, Hogan and Retzel (1995) note that in a bibliography compiled by Charles Magel in 1989, there were only 107 items published from ancient times up to 1974 but more than double that number between 1975 and 1988. I could find only about 100 books, articles, and reports that were first published between 1900 and the 1960s, which is just over 10 per cent of the entire bibliography. Scholarly output increased dramatically in the 1980s (Chase 1990), and as noted in the introductory chapter, there

is evidence since then of increasing interest in animal rights in a number of academic disciplines. These data suggest a decline in interest during the first half of the twentieth century on animal welfare issues as measured by publishing output.

2. Jasper and Nelkin (1992:177) provide a graph of paperback sales and books in print as an indicator of the growth of the modern movement during the 1970s and 1980s, noting that the number of books in print by 1988 had climbed to over a quarter of a million.

Chapter 2

Profile of a Social Movement

If you want drama, get a movement; if you want results, get an organization.
—Saul Alinsky, *Rules for Radicals*

In this chapter, I discuss some of the key social-scientific concepts used in the study and the main animal protection organisations in the three case study countries, Australia, the United Kingdom, and the United States. For many ordinary people, animal rights remains a curious abstraction, and indeed most writers perceive the animal rights movement as "a movement of ideas," to use Finsen and Finsen's (1993) apt phrase. But there is a danger that in privileging the ideas of the animal liberation philosophers, we ignore the ordinary grassroots activists and organisational advocates whose voices are too easily overlooked in the illustrious company of the Regans and Singers in the movement. The purpose of the book is to show how the ideas of the movement are also the ideas of the individuals and groups whose campaigns in the streets and in the suites constitute the core of the movement. This chapter provides a sociological perspective on social movement activism and advocacy, and introduces the organizations that form the basis for the study of the animal movement in the three case study countries.

ANIMAL ACTIVISTS AND ADVOCATES

Compared to their Australian counterparts, animal protectionists in the United Kingdom and in the United States are more easily categorised as either activists or advocates. Because of the different political systems in these countries, access to decision makers determines the style of animal protectionism that is most effective. While both countries have essentially a two-party system, the American system is more amenable to the lobbying of interest groups than is the United Kingdom system, which relies much more on policy communities and issue networks (Smith 1993; Ruedig 1995:220; Garner 1995). As access to United Kingdom decision makers is more restricted than in the United States, agents of social change are much more inclined to take to the streets in Britain to get heard. Furthermore, the tradition of social protest in the United Kingdom, such as the trade union movement, is more firmly established than it is in the United States.

Zald suggests that political structures in the United States are sufficiently open and consultative which make it easy for both pure grassroots movements and pure national lobbying organizations to gain legitimate status (1988). Even so, grassroots organisations as defined by Rucht are much less prominent in the Washington lobby belt and, hence, have a relatively low public visibility and profile (1996:188). As we shall see in Part II, this is not the case in the United Kingdom, where grassroots activists have orchestrated some of the most important protests in the animal protection movement.

In contrast to both these political milieux, the political system in Australia tends to be relatively more egalitarian, in that small organisations and even individuals acting on their own often do form, for tactical reasons, alliances or mergers that are unlikely in the United States. In a country of only nineteen million people, Australian social movement activists and advocates are more likely to need each other than those in the United States and the United Kingdom do, where social movement sectors are much larger. As noted in the previous chapter, the need for critical mass compelled Australia's animal protectionists and conservationists to form an uneasy alliance in the nineteenth century. Under such conditions, both activists and advocates in movements for social change in Australia operate in an arena that does not necessarily favor one form of advocacy/activism over the other. While some agents of social change might dispute this, there is little doubt that alternative ideas and progressive movements for social change in Australia are likely to be less dependent on the financial resources or institutional clout of professional interest groups than they are in either the United States or Britain. Organisationally, the animal protection movement in Australia is significantly smaller than it is in these countries, although there is evidence that the movement has a substantial grassroots following (Munro 1996b).

The interviewees in the study all see themselves as animal protectionists, a term that is comprehensive enough to include the philosophical differences among animal welfare, animal liberation, and animal rights. The term also includes antivivisectionists and those who described themselves as activists, advocates, and supporters in the animal movement. This study distinguishes between grassroots activists in the streets and organisational advocates in the suites, since the individuals either designated themselves as one or the other or fitted the category by the role they played in the movement. Those who are employed in animal welfare organisations as paid employees can be more accurately described as advocates, especially since their position in charitable societies prohibits their involvement in direct action of a political kind.[1] By contrast, those who are involved on the fringes of social movement organisations (SMOs), or are unpaid volunteers, are generally involved in the kind of direct action that is the hallmark of the activist. However, as I pointed out in the introductory chapter, the designations between activist and advocate are not always clear-cut for some organisations and are even less so for some individuals interviewed for this study. Table 2.1 lists the interviewees and their organisational affiliations where appropriate. Real names are given in full, and pseudonyms are used for those who preferred anonymity.

The Expressive and the Instrumental Sides of Animal Protection

Staggenborg (1988:599) argues that formalized SMOs tend to engage in institutionalized tactics, such as legislative lobbying, in contrast to the direct action that grassroots activists favour. She refers to another important difference in styles of social movement advocacy and activism. Formalized SMOs need to stay the distance, particularly in unfavourable environmental conditions when it is difficult to mobilise constituents. Thus, while grassroots activists initiated and executed a hunger strike in London, it was left to SMOs like Animal Aid and CIWF to continue the protest from the suites after the hunger strikers had gone home. And as Staggenborg's study makes clear, formalized SMOs have the organisational structures and professional staff to sustain themselves by mobilising resources, ensuring that the movement's issues are kept in the public eye during lulls in movement activity. As explained in the previous chapter, the animal movement's campaigns were held in "abeyance" (Taylor 1989) for much of the first half of the nineteenth century by means of such networks.

As I will show in Chapter 6, animal protection work is demanding and emotionally draining, particularly at times when the movement is engaged in a major campaign, such as the live animal trade issue during the mid-1990s in the United Kingdom. It is then that the potential for conflict between volunteers and professionals is likely to be greatest. Ironically, it is also the time when both approaches, direct action and advocacy, need to be in harmony. One pro-choice individual put the case for balance as follows:

> Without the grassroots people, I think things would be overstructured; with just the grassroots people, there's too much burnout among them. The staffers tend to last a lot longer (Staggenborg 1988:603).

Adele Douglass, executive director of the American Humane Association, was asked to describe what she saw as the most important dimensions of her job.

Table 2.1
Animal Protectionists Interviewed in the Case Study Countries

Activists

Australia (12)	*UK (9)*	*USA (5)*
Casey	*Andrew Tyler (Animal Aid)	Alan
*Gaynor	Gary	*Christine Stevens (AWI)
*Jim Roberts (AL Vic)	James	*Scott Williams (FARM)
*Laurie Levy (CADS)	Joan Court (ARC)	*Wayne Pacelle (HSUS)
*Leslie	*Mike Huskisson (ACIG)	Tina
*Lisa	Milly	
Owen	*Morry	
*Patty Mark (AL Vic)	Phyllis	
Rhett	*Tim O'Brien (CIWF)	
*Rheya Linden (AL Vic)		
*Roger		
*Sherry		

*Denotes those for whom the designation "activist/advocate" is more accurate.

Advocates

Australia (10)	*UK (5)*	*USA (12)*
Al	Barbara	Adele Douglass (AHA)
Carole de Fraga (WSPA)	Collette Kase (NCDL)	Ann Sparks (AHA)
Elizabeth Ahlston (AAHR)	John Bryant (LACS)	Carter Luke (MSPCA)
Joan Papayanni (WLPA)	Joyce d'Silva (CIWF)	Cathy Liss (AWI)
Glenys Oogjes (ANZFAS)	Kaye	Gus Thornton (MSPCA)
Jenny Talbot (Proj Jonah)		Holly Hazard (DDAL)
Margaret Bowman (WLPA)		Pat Reilly (MSPCA)
Mark Berriman (AVS)		Stephanie Roderick (WSPA)
Sid		Tamara Hamilton (HSUS)
Sue		Tom Regan
		Vicky Kysar (WSPA)
		**Robert Ferris (DoW)

**Denotes that this interviewee preferred the designation "supporter."

Her reply was fairly typical of those who were paid professionals in animal protection organisations in the three case study countries:

> *It's challenging, I like the strategy. I absolutely love the strategy of figuring out how to do something. I like the road blocks that are thrown in, I mean it makes it more challenging of how to get around those and when you come across people that are very hostile how to not offend them and get around them. I guess I like the politics of it.* (Interview 1996)

Table 2.2
Cross-National Campaign Orientations in Animal Advocacy/Activism

UK	USA	Australia
grassroots activism	organisational advocacy	grassroots activism
>	>	+
organisational advocacy	grassroots activism	organisational advocacy

Douglass worked in congress for most of the 1980s, and at one time was a representative from New York City. One of the issues she dealt with was animal welfare, since no one liked dealing with it because of the so-called "crazy constituents." Douglass enjoyed the work even though it was stressful at times:

I enjoyed the issue, I have to be honest with you. I have always grown up with animals, I love animals, and I have a very strong ethic of never harming any other living thing. So I did the animal issues and dealt with the "crazy constituents" which was humorous sometimes and sometimes very trying. (Interview 1996)

The "crazy constituents" term refers to the grassroots activists whose expressive style of activism politicians would prefer to ignore. "Emotional" defenders of nature—"animal libbers," "econuts," and "bunny- and tree-huggers"—are at times kept at a safe distance by their more moderate colleagues in the suites as well. In an analysis of the environmental movement, Eyerman and Jamison refer to the tensions between the amateurism of "grassroots epistemology" and the professionalism in which "the organization has become institutionalized as a tool or weapon for the advance of specific goals" (1989:103). Expressive concerns motivate the amateurs and more instrumental goals motivate the professionals.

ACTIVISM AND ADVOCACY STRATEGIES IN THREE COUNTRIES

While there is often overlap between the activists and advocates in various campaigns, the distinction is a useful way of organising the data. At the broadest level of analysis, the three countries in the study are different in the animal protection strategies that they adopt, although the tactics used are often similar. The cross-national differences are summarised in Table 2.2.

As Table 2.2 shows, grassroots activism is more common than organisational advocacy in the United Kingdom while the reverse is true for the United States. An explanation for this ideal-typical schema lies within the political systems for each country. Garner argues that in Britain, regular and effective access to decision makers by animal protectionists has proven to be largely elusive (1995:

56). In the United States, on the other hand, the political system is more open to the influence of animal protection lobby groups, especially those with offices in Washington D.C. The relationship between animal protection SMOs and the state in Australia can roughly be described as falling between the United Kingdom and the United States models, both of which have larger and more professionalized animal protection SMOs.

Organisational Tactics and Politics of Animal Protection

It is important to point out that the term "social movement organisation" (SMO) is used by sociologists to denote what political scientists refer to as pressure groups, lobby groups, or interest groups. Jennett and Stewart suggest that social movements differ from these groups and political parties in that social movements seek the moral high ground and to change people's hearts and minds (1989:1). Rootes draws a further distinction when he suggests that there is something utopian about social movements, which no observer seeking to understand a particular movement should overlook (1984:6). Social movement organisations involved in animal protection are nonetheless engaged in *realpolitik* and consequently need to temper the utopianism of the grassroots by focussing on winnable issues. They do this, often in collaboration with their grassroots supporters, by adopting strategies appropriate to their identities. Some SMOs work within the system, such as ANZFAS in Australia, the American Humane Association in the United States, and the League Against Cruel Sports in the United Kingdom, while the more radical agendas and identities of others, like the Animal Liberation Front and the Hunt Saboteurs Association, place themselves outside or on the margins of the political system.

Differences in ease of access to political decision makers means differences in the strategic and tactical repertoires that activists and advocates in the animal movement are likely to use. As Ruedig points out, a system in which one or two political parties dominate would usually have the effect both of moderating the actions of groups wanting to join the system and of alienating and radicalizing excluded groups (1995:220). Animal protection SMOs that enjoy good relations with decision makers are unlikely to risk their position of influence by engaging in covert, illegal, or violent actions. On the other hand, groups that have no stake in the political process are more inclined to follow this radical orientation, often out of a sense of frustration.

Although movement leaders and mainstream supporters alike denounce illegal and violent acts perpetrated in the cause of animal protection, they have a number of unintended consequences for the movement. Some movement watchers have suggested that the extreme actions of radicals can have the effect of legitimating and strengthening the bargaining position of the moderates who are made to appear reasonable by comparison (Scarce 1990:6–7; McAdam et al. 1988:718–9). The more radical elements in the movement then are compelled to form their own campaign organizations because they cannot abide the "respectability" of the movement or its newfound appeal to traditional adversaries. The history of the animal protection movement is replete with instances of activists building alternative organizations that they see as more effective than their forerunners were in challenging their opponents.[2]

Another response to extremism can simply be described as the defensive theory of social movement politics, whereby countermovement organisations mobilise to defend their interests in the wake of sustained attacks. It is not unreasonable to argue that one measure of a social movement's success is the intensity of opposition to it. Any social movement that is taken seriously and challenges existing codes, moral values, or vested interests invariably inspires the emergence of robust countermovements and sometimes police surveillance. Both forms of defence have emerged to counter the claims and activities of the animal liberation movement in all three case study countries (Munro 1999b).

Turner and Killian suggest a number of tactical mechanisms that are relevant to how animal protectionists in the streets and in the suites attempt to influence different audiences (1987). All four mechanisms—persuading, facilitating, bargaining, and coercing—have been used at one time or another by activists and advocates in their campaigns on behalf of animals. These tactical mechanisms can best be thought of as a continuum, with persuasion as the most moderate tactic at one end and the more direct confrontational tactic of coercion at the other end. Not surprisingly, persuading, facilitating, and bargaining tend to be the preferred tactics of organisational advocates in the suites, while coercive tactics are usually more commonly observed in grassroots activist campaigns.

Persuasion, involving the use of strictly symbolic manipulation and the raising of issue consciousness, is one of the most important ways in which ideology is produced and continuously modified (Turner & Killian 1987:297–98). For organisational advocates, persuasion usually takes the form of education campaigns, typically via their own print media (e.g., brochures, glossy magazines, and the like). Although consciousness raising in the environmental movement has been derided as "social change through banner hanging" (Wapner 1995), it is nonetheless an important tactic in the animal rights movement for changing the way people think about animals. Wapner uses the example of the whale to show how effective symbolic protests such as "banner hanging" have been for Greenpeace. In the 1960s, the United States Navy and Air Force used whales for target practice; in the 1980s, an international effort costing $5 million was organised to save three whales trapped in the Alaskan ice (Day 1987:157). Numerous examples of the use of persuasive communication as a tactic for changing people's sensibilities will be referred to throughout this book.

Likewise, there are many instances in which coercive tactics of various kinds have been deployed, particularly by grassroots activists, to achieve improvements in animal welfare. These range from the use of "nuisance" tactics to more disruptive methods, including the violent actions of extreme animal rights activists. Needless to say, virtually every animal protection SMO eschews these forms of direct action, although much depends on how direct action is defined.

Rescuing an animal from the clutches of a hunter would constitute a form of direct action that virtually every self-respecting animal advocate would support; on the other hand, liberating an animal illegally from a laboratory or factory farm would not be encouraged. Tensions between advocates and activists often arise over tactics, as we will see in the various campaigns described in the study.

Grassroots activists are more inclined to embrace the symbolic and expressive, with little concern for the more pragmatic evaluation of long-term strategic planning that animal advocates and lobbyists favour. Turner and Killian note that an undisciplined grassroots movement employs many tactics more expressively than strategically (1987:301). Evidence of spontaneity is plentiful among grassroots animal protectionists in this book, just as there is an observable preference for the strategic over the expressive among advocates in the suites. Yet these distinctions are never cut-and-dried. For example, the tactical mechanisms of bargaining and facilitating, which theoretically belong in the suites and offices of the professional lobbyist, are sometimes employed by grassroots activists in forming coalitions with other groups or in facilitating a target group's support of the movement (e.g., the unlikely alliance between animal liberationists and Australian meat workers over the issue of live animal exports). As we will see in the various case studies and exemplary animal protection campaigns described in this book, activists and advocates make their own history without recourse to sociological prescriptions of how things are meant to be.

CARING SLEUTHS

There are a number of ways of defining activism in the social movement literature, ranging from generic social movement activism to specific animal rights activism. The uniqueness of animal rights activism is partly captured by Shapiro's five propositions that he claims define the prototypical animal rights activist:

> An animal rights activist is an individual whose primary concern is caring about animals; who is primed to see suffering in animals; who aggressively seeks out and skillfully investigates situations in which animals are suffering; and for whom such caring, seeing and seeking become pervasive aspects of daily life, embodied in his or her lifestyle; tensions between the apparent contradiction between an attitude of caring and the aggressive exposure of human-oriented animal suffering are resolved in one of several ways: embracing, suppressing, or losing touch with the caring. (1994:148)

Oliver and Marwell provide a generic definition in their description of social movement activists as "people who care enough about some issues that they are prepared to incur significant costs and act to achieve their goals" (1992:252). This generic definition has the advantage of breadth and inclusiveness, and it incorporates Shapiro's ingredients of care, action, and costs (or "tensions" in his terminology). Shapiro's definition has the advantage of specificity but suffers from being overly individualistic and prescriptive. Nonetheless, it is an accurate statement of how Shapiro's activists experienced activism at a psychological level since "caring, seeing and seeking" were evidently part of their daily lives. The notion of "the caring sleuth" is also close to what many people inside and outside of the movement perceive as the prototypical animal rights activist, namely an individual who is prepared to *do* something, no matter what the cost, about animal suffering. While I did not test Shapiro's propositions in any rigorous way, they were informally used to find out more about an individual activist's motivations and what meanings he or she attached to animal rights activism.

Profile of Animal Protection Organisations in Australia, the United Kingdom and the United States

It is notoriously difficult to calculate the size of the animal movement worldwide or even in the three case study countries. Estimates of the movement worldwide range from ten to fifteen million, although Singer, at a time when the movement had peaked, thought that these figures were unrealistically large (Singer 1985:47). The best that we can do is to provide rough estimates of the membership of organisations and groups that belong to the movement and publish their activities in directories and the like.[3] There is little doubt that the United States has the largest number of animal advocates, with over four hundred advocacy groups, a total membership of ten million, and an annual income of $50 million (Blum 1994:113–4).

More important than the size of membership of the movement however, is the range and diversity of grassroots and advocacy organizations. Table 2.3 shows the organizations that I visited for this study. In most cases, the designation "advocacy" or "activist" organization was disclosed in the interviews with staff members or was suggested in the literature provided by the organization. These should be interpreted as broadly accurate descriptions rather than as rigid definitions of the organization's ideological or programmatic orientation. For instance, staff at the Animal Welfare Institute (AWI) and FARM tend to see themselves as activists, although both groups are headquartered in Washington D.C. and spend much of their time lobbying Congress from the suites.

As shown in Table 2.3, half of the SMOs have an activist orientation and half are characterised by an advocacy orientation. While in no way comprehensive, I believe that they are broadly representative of the animal movement in the respective countries.

Some other organizations not listed in Table 2.4 but referred to in the book include People for the Ethical Treatment of Animals (PETA), the Hunt Saboteurs Association (HSA), and the Royal Society for the Prevention of Cruelty to Animals (RSPCA). Because these organizations are among the best known in the animal movement and well represented in the literature, I decided not to include them in the interviews. However, several individuals indicated that they

Table 2.3
Animal Activist and Advocacy Organisations in Australia, the UK, and the USA

Activist Organisations

• *In the United Kingdom*: Animal Aid, Animal Rights Cambridge (ARC), Brightlingsea Against Live Exports (BALE), Animal Cruelty Investigation Group (ACIG);

• *In the United States*: Farm Animal Reform Movement (FARM), the Animal Welfare Institute (AWI);

• *In Australia*: Animal Liberation (Victoria, NSW), the Coalition Against Duck Shooting (CADS), Mountain Residents for Animal Rights (MRAR), the Guardians, Project Jonah.

Advocacy Organisations

• *In the United Kingdom*: Compassion in World Farming (CIWF), the League Against Cruel Sports (LACS), the National Canine Defence League (NCDL), the British Union for the Abolition of Vivisection (BUAV);

• *In the United States*: The American Humane Association (AHA), the Humane Society of the United States (HSUS), Defenders of Wildlife, the Massachusetts Society for the Prevention of Cruelty to Animals (MSPCA), the Doris Day Animal League (DDAL):

• *In Australia*: The Australian and New Zealand Federation of Animal Societies (ANZFAS), Animals Australia; the Australian Association of Humane Research (AAHR); the Australian Vegetarian Society (AVS); the World League for the Projection of Animals (WLPA).

belonged, at some stage, to one or more of these well-known groups. It was quite common for the staff of some of the larger SMOs to have done their apprenticeship in the HSA or PETA.

All of the selected organizations are involved in animal protection, either at the grassroots or organizational level or both. As animal protectionists, they can be further designated as animal welfarists, animal liberationists, and animal rightists. In deciding on these categories, I was guided by the terminology used in the literature. "Animal welfare" is typically associated with the idea of kindness to animals, best exemplified by the RSPCA. The term "animal liberation" is more problematic since it is often associated in the public's mind—in the United Kingdom and the United States but not in Australia—with the Animal Liberation Front. In Australia, the term "animal liberation" is relatively benign, even respectable because of its association with Singer's name. "Animal liberation," as used in this study then, refers to the pragmatic, nonviolent activities of the movement as espoused in Singer's book of that name (1975, 1990). Finally, I use the term "animal rights" as it was intended by Regan in *The Case for Animal Rights* (1984). It means activities designed to abolish, rather than

Table 2.4
Programmatic Orientation of Select Animal Protection Organisations in the Case Study Countries

Animal welfare	Animal liberation	Animal rights
(RSPCA)MSCPA/WSPA	AWI	Animal Aid
NCDL	Animal Lib (Vic)	ARC
Defenders of Wildlife	Animal Lib (NSW)	ACIG
	Project Jonah	FARM
	CIWF	CADS
	LACS	Guardians
	AHA	AAHR
	HSUS	BUAV
	ANZFAS	MRAR
	WLPA	(PETA)
	BALE	(HSA)
	AVS	
	DDAL	

reform, human (ab)uses of animals. Again, these designations, as shown in Table 2.4, were provided in the interviews or gleaned from the organization's literature and are meant to be broadly descriptive rather than definitive.

The RSPCA is usually seen as quintessentially animal welfarist in its activities. The World Society for the Protection of Animals (WSPA) is much more difficult to categorise; although it is a partner of the RSPCA in the United Kingdom and Australia and of the Massachusetts Society for the Prevention of Cruelty to Animals (MSPCA) in the United States, its field workers are often involved in quite dangerous work, as I explain in Chapter 10. At the other end of the continuum, groups like PETA and HSA see themselves as abolitionists who want humans to leave animals alone. These designations are open to question, and some movement commentators would no doubt challenge them. For instance, Gary Francione (1996) categorises most organisations that claim to follow a rights orientation as part of a "new welfarism" that is far removed from the animal rights position he believes the movement ought to embrace. Nonetheless, I was interested in how the organisations portrayed themselves in their promotional literature and how movement adherents saw them.

What all of the above welfare, liberation, and rights groups have in common is an aversion to violence. Just as they are opposed to the violence that humans do to animals, they reject the use of violence against animal exploiters perpetrated by animal defenders on behalf of animals. Abolitionists who follow Regan's lead, in common with all the groups listed above, are totally opposed to the use of violent or illegal means to achieve their ends. Extremist groups outside the mainstream movement organizations listed above were not included in the interviews for reasons that had to do with gaining approval for the study.

I attempted to include large and small organizations that represented the var-

Table 2.5
Animal Protection Organisations by Issue or Focus

- *Multi-issue groups*: AA, ARC, AWI, HSUS, DDAL, AHA, ANZFAS, WLPA, WSPA
- *Specialist or focussed campaigns*: ACIG (mainly cruelty to lab or wild animals), **Farm animals**—Animal Liberation in Australia, FARM, CIWF, MRAR, BALE; **Domestic animals**—RSPCA/MSPCA, NCDL.
- *Single-issue groups*: **Wild animals**—CADS, LACS, Project Jonah, Defenders of Wildlife; **Laboratory animals**—BUAV, AAHR, Guardians.

ious issues of the animal movement in all three countries. The two dozen organizations fell roughly into three categories, as shown in Table 2.5.

The multi-issue groups include a variety of animal protection SMOs, ranging from the large, wealthy, national Humane Society of the United States (HSUS) to local, grassroots, direct-action groups like Animal Rights Cambridge (ARC). Multi-issue SMOs almost always have divisions that include the three big animal protection campaigns against factory farming, vivisection, and hunting and trapping—as well as a host of other issues such as circuses, zoos, rodeos, steeple jumping competitions, companion animals, film making, and so on. Wealthier organizations, like HSUS and the American Humane Association (AHA), also sponsor humanitarian education and antiviolence campaigns in addition to these conventional animal protection concerns. Most of the SMOs do not have the resources to mount several campaigns simultaneously and so focus on a particular issue such as wild, domestic, laboratory, and farm animals.

This division of labour also helps the movement as a whole avoid too much duplication in the animal issues marketplace. In some cases, the name of the SMO is not necessarily a good guide to the issues and campaigns. For example, farm animals have tended to be the preoccupation of Animal Liberation groups in Australia, although the issues at different times have varied from branch to branch. And while the American-based Farm Animal Reform Movement (FARM) is widely known for its campaigns against intensive farming, it has recently come to be more like a vegetarian society with its adoption of an environmental/vegetarian focus. Similarly, Mountain Residents for Animal Rights (MRAR), a small grassroots group near Sydney, strictly speaking does not pursue an animal rights philosophy and is closer ideologically and programmatically to Animal Liberation as it exists in Australia.

Animal protection SMOs, however, are usually meticulous in their use of logos, as the name of an organization can quite literally determine its credibility or fund-raising capacity. Compassion in World Farming (CIWF), the League Against Cruel Sports (LACS), and the British Union for the Abolition of Vivisection (BUAV), while not household names in the United Kingdom, are likely to have more public appeal than No More Victims, Pet Sitters International, or the Society of Kind Understanding for Not Killing Skunks (SKUNKS), which are some of the groups listed in the American Animal Rights Directory. (*The Animals' Agenda* July-August, 1996). Single-issue organizations often use ingenious acronyms such as PAWS (Performing Animals Welfare Society), CADS (Coalition Against Duck Shooting), and the ad hoc group BALE (Brightlingsea Against Animal Exports), which campaigned against the trade in sheep and other animals in the United Kingdom.

Just as there are many different SMOs that make up the worldwide animal movement, there are several schools of thought and philosophies that constitute the movement's ideology as a campaign to change the way humans think of and treat animals. Elsewhere in the book, I cite evidence of a strong ideological consensus in the movement for what animal protectionists perceive as the worst forms of cruelty perpetrated against animals and how these abuses might best be ended. For the moment, however, the importance of philosophical ideas in the movement's development is discussed in the next chapter.

NOTES

1. Animal protection SMOs in all three countries operate as nonprofit, charitable organisations that enjoy tax-free status or substantial tax deductions. The term "nonprofit" is misleading, however, since the organisations are only restricted from distributing their profits, not from using their income to expand operations and infrastructure, hire more personnel, and the like. Having charitable status does, of course, restrict the organisation's freedom to agitate against government policy. Thus, an organisation like the RSPCA in the United Kingdom has always had to tread a careful path on the issue of fox hunting, a popular pastime for many members of the British House of Lords, not to mention the Royal Family.

2. The League Against Cruel Sports's defection from the RSPCA, and later the Hunt Saboteurs Association's break with the league, is a classic instance of this phenomenon. Similar high-profile "divorces" have occurred between NAVS and the BUAV in Britain and between the AHA and the HSUS in America.

3. Thus, one directory of American animal groups published in *The Animals' Agenda* (July-August, 1996) lists over 300 groups, ranging from large, well-known SMOs like PETA to small outfits that are little more than letterheads. A similar list of Australian groups published by ANZFAS puts the number of animal welfare/rights groups at just over 90, although many of these groups would be in the letterhead category. The membership of the animal protection movement in Australia is roughly in the vicinity of 127,000, with ANZFAS member societies and the RSPCA providing about half the total number. The RSPCA in the United Kingdom has about 500,000 supporters in 200 branches in England and Wales and 178 affiliated groups worldwide (*The Economist* 19 August, 1995a:17). By comparison, Australia's RSPCA has approximately 30,000 supporters. Another source lists a total of 62 organisations in the United States and 40 in the United Kingdom, covering 11 different categories of animal advocates (excluding vegan and vegetarian groups): multipurpose, laboratory animals, farm animals, wildlife, sport and entertainment, companion animals, marine life, specific animals, professional organisations, special interest groups, and legislative organisations.

Chapter 3

The Compassionate Beast: A Movement of Hearts and Minds

Philosophy can lead the mind to water but only emotion can make it drink.
—Tom Regan, animal rights philosopher

Animals are not affected by how we feel but by what we do.
—John Webster, professor of animal husbandry

It will be clear by now that this study is not about "animals" or "rights" per se. Rather, it is concerned with a social movement that seeks to change the way human beings treat animals.

In this chapter, the work of some of the most important critics of the animal movement are outlined and evaluated. Like these scholars, I have found it difficult to remain detached, perhaps because of the nature of the subject, the complexity of which is so accurately captured in the epigrams at the head of this chapter. This book is essentially about these two ideas expressed by Regan, the animal rights philosopher, and Webster, the practising animal scientist—what motivates animal protectionists in their campaigns on behalf of animals, and what are the results of the strategies and tactics that they use?

Activists and advocates invariably perceive animals as innocent victims of human cruelty. A desire to end what they see as an outrageous injustice perpetrated by human beings against animals drives their campaigns. The animal movement has been described as "a moral crusade" in which campaigners use emotions, moral shocks, and atrocity stories about "the suffering of innocents" to mobilise public support for their cause (Jasper & Nelkin 1992; Jasper &

Poulsen 1995; Groves 1995). For Finsen and Finsen (1994), the animal movement is the quintessential movement for social justice, while to Tester (1991), it is not about animal rights at all but rather about how adherents define their humanity. Conflicting interpretations are a constant theme in the sociological literature on the animal movement, just as they are in the work of the philosophers who have shaped the movement from the outset.

A MOVEMENT OF IDEAS

Philosophers, according to Jasper and Nelkin are the midwives of the animal rights movement (1992). No serious book on the subject of animal rights will fail to include the contributions of Singer, Regan, Clark and other movement philosophers, if only to clarify the terminology used in liberation discourse. As we will see elsewhere in the book, there are many who find the notion of animal rights ridiculous or "desperate," to use Mary Midgley's description of "rights talk." As a prominent and respected animal advocate, Midgley believes that the word was already in serious trouble well before animals were added to its worries (1983:61). Even so, while philosophers continue to debate the applicability of attributing rights to animals, movement activists widely use the term "animal rights" as a campaign slogan.

Magel suggests that "animal rights" is essentially an ethical term and therefore an issue for philosophers to resolve (1989:x), while Clark maintains that the role of the moral philosopher is suggestive rather than prescriptive: "Our value is not that we have answers, but that we have questions, and some knowledge of the puzzles created by attempted answers of the past" (Clark 1997:113).

Some of the questions raised by a number of leading philosophers on both sides of the animal rights issue will be examined in the remainder of this chapter. In *The Liberation Debate: Rights at Issue*, edited by Leahy and Cohn-Sherbok (1996), the editors include animal liberation alongside liberation theology and women, black, gay, and children liberation as their focus of study. Michael Leahy, although himself a strong critic of the animal movement's ideology, suggests that the term "liberation" is most apposite in the context of animals since of all these disadvantaged groups, animals—together with the oppressed masses of the Third World and, to a lesser extent, children—are most in need of liberation. According to Leahy, the term conjures an image of people in chains, is more emotive, and carries greater rhetorical force than pedantic and legalistic "rights talk."

Singer's *Animal Liberation* (1975) has often incorrectly been characterised as an argument for animals' rights, a mislabelling that Singer is happy to accept since the term is a convenient political slogan. And among supporters and critics of the animal movement, the terms "liberation," "welfare," and "rights" are typically used interchangeably as they are understood as "distinctions without differences," to use Wayne Pacelle's phrase (Francione 1996:227). Few people associated with the movement, apart from moral philosophers and academic writers, most notably Francione, are interested in the distinctions that the terms represent. Nonetheless, what philosophers mean when they describe the different approaches to animal protectionism is important since terms like "animal rights" can lose all coherence without some common agreement as to their intended meaning. And in a field in which there is often more heat than light, words are important. I will therefore outline some key ideas of some prominent philosophers on both sides of the animal liberation debate.

Singer/Regan and Leahy/R. G. Frey

The Australian philosopher Singer has been called the most effective philosopher alive (Appleyard 1995), while critics have denounced his work as "a clear illustration of the moral bankruptcy and intellectual impoverishment of the animal liberation philosophy" (Nicoll, Russell & Lau 1992:60). Similarly, Regan's *The Case for Animal Rights* (1984) is cited by some as the most rigorous and philosophically sophisticated argument for the extension of rights to animals (Benton & Redfearn 1996:50), while others like Steven Rose have denounced Regan as "muddle-headed and intellectually dishonest" (Benton 1992:118). Yet despite these contradictory reviews, every serious book on the animal movement devotes a section or two to the ideas of Singer and Regan (pronounced as in "vegan"). Their most important contributions in guiding us as to how we should treat animals are outlined below. The Singer/Regan pro-animal liberation perspective will be described alongside the critique by Leahy, who utilises R. G. Frey's case against animal liberation, primarily in *Interests and Rights* (1980), to craft his own rebuttal of Singer/Regan and other animal liberationists. The arguments for and against liberation will then be evaluated within the context of the main arenas of movement activism—factory farming, animal experimentation, and blood sports—and will conclude with Alan Wolfe's (1993) defense of anthropocentrism. That more space in this chapter is given over to the animal movement's critics can be justified on the ground that the bulk of *Compassionate Beasts* focuses on the views of movement supporters, advocates and activists.

The Singer/Regan Case for Animal Interests/Rights

Although commentators often emphasise the differences between their philosophical positions, Singer and Regan are on the same side in that they both want people to respect animals as sentient beings that deserve to be treated compassionately, rather than as if they do not matter. Both philosophers have published, individually and as a duo, and agree that the worst forms of cruelty humans perpetrate against animals occur on factory farms, in laboratories, and in hunting and trapping. But their differences are important for those who are guided by what they have to say about the way we should treat animals.

Interests

Singer argues that when we mistreat animals by eating them, dissecting them, and killing them for sport, we are guilty of speciesism, an act of discrimination that he claims is on par with racism and sexism. According to Singer, race, sex, and species membership are morally irrelevant criteria upon which to discriminate against other beings. Like the social movements that seek to emancipate oppressed racial/ethnic peoples and women, the animal liberation movement wants to liberate animals that humans exploit and abuse.

Antispeciesism requires that we give equal consideration to the interests of all sentient beings. Pain is pain whether suffered by a boy or a dog, and it is in the interests of both to avoid it. Most animals and all humans, unlike trees and rocks, have the capacity for experiencing pain and pleasure. It is not animal rights that Singer advocates, but rather the animal's interest in avoiding pain. Thus, a Singerian would argue that the boy's interests and the dog's deserve equal consideration, but not necessarily equal treatment. A dog might need to be put down if it attacked a boy; but not vice versa. But a dog should not be required to suffer so as to satisfy the curiosity of a research scientist or to forfeit its life to facilitate some trivial research project. However, replace the dog with a rodent and most people would probably be less enthusiastic about supporting the proposition. As a utilitarian, Singer accepts that animals' lives may need to be sacrificed for a greater good, but he urges us to weigh carefully the animals' lives against the supposed benefits. This is the theory of utilitarianism, which maintains that an act is right if, and only if, in the absence of a better alternative, the amount of good that the act causes outweighs the amount of bad. For instance, that millions of animals die to satisfy our taste for meat, fur, cosmetics, or fun is, to the utilitarian, morally wrong.

To Singer, it is the capacity for suffering, not species membership, that dictates how we should treat animals. A human equipped with the powers of imagination and diagnosed as terminally ill suffers in a way that no animal can. By the same token, battery hens, tethered sows, and caged lions cannot comprehend the reason for their confinement, and so it may be that they endure greater stress than the confined convict or slave who at least understands the reason for the incarceration. Thus, antispeciesism requires that we consider the interests of all sentient beings equally so that it is not just humans or charismatic megafauna—whales, white rhinos, and pandas—who are the recipients of moral consideration, but the millions of laboratory and food animals that are Singer's principal concern: "As long as we remember that we should give the same respect to the lives of animals as we give to the lives of those humans at a similar level, we shall not go far wrong" (Singer 1975:24). As a rule of thumb, this means that if we object to experiments being conducted on a brain-damaged human, we should logically find similar experiments on a healthy animal morally reprehensible.

Singer's utilitarianism allows for the humane killing of sentient beings under certain circumstances where the positive consequences of the act outstrip the negative. Thus, the deaths of a number of animals in an experiment that leads to the discovery of a vaccine that in due course saves both animal and human lives may, according to a utilitarian calculus, be justified. But in *Animal Liberation* (1975), there would seem to be no justification for the killing of millions of animals annually on factory farms and in laboratories. Killing animals for food, when there is a vegetarian alternative, is indefensible to Singer. With animal experimentation, each case would have to be decided on its merits, since there are instances when the aggregate good of humans and animals may be served by the sacrifice of individual animals. This is the kind of pragmatism that abolitionists in the movement reject since the utilitarianism cost/benefit calculus allows individuals to be harmed for a greater good, a proposition abhorrent to the likes of the American philosopher Regan.

Rights

Regan's uncompromising abolitionist stance is, by definition, at odds with Singer's pragmatism regarding how we should treat animals. Regan finds Singer's utilitarianism inadequate for defending the rights of animals to be left alone. They are in Regan's terminology "subjects-of-a-life," which means that they have inherent value that humans should respect. Inherent value is something that Regan argues all individual human and nonhuman animals have equally. It is inherent value that is the basis for the claim of rights. If animals have rights, the most basic is the right to bodily integrity, which an activist interviewed for this book graphically described in the case of battery hens as "*the right to own a beak.*" To Regan, animal rights means an animal has a right to respectful treatment. However, this is more than the utilitarian principle of equal consideration of the animal's interests, for rights explicitly mean the right not to be harmed. If the rights of animals are to be taken seriously, no practice—experimentation, commercial agriculture, hunting and trapping—that results in the suffering or death of an animal can be justified. Regan and his animal rights followers therefore call for nothing less than the abolition of these "abuses," irrespective of whatever benefit they may be to other animals or humans. Animal exploitation, like slavery or the oppression of powerless people, is morally wrong. According to this view, zoos, circuses, and even pet keeping should be banned.

For freedom-loving people, the doctrine of individual rights has enormous appeal since it supports the idea that the individual's liberty cannot be sacrificed for the common good. Yet "rights talk" is fraught with moral pitfalls and can be easily exploited by either left-wingers or right-wingers to argue a convincing case, for instance, about the virtues of big or small government. The Right can put a strong case for keeping government out of people's private lives—the right to be left alone, including the provision of welfare—while the Left can claim that the failure of government to deliver adequate material assistance to the needy is a violation of human rights. The issue of rights is even more complicated, some would say incomprehensible, when it is applied to the lives of animals. "Rights talk" readily becomes a weapon that critics and sceptics of animal liberation use to undermine the defence of animals' rights (see, for instance, the Senate Select Committee's report in the next chapter). It is too easy, even for those who are sympathetic to the animal cause, to claim that just as people do not have a right not to be eaten by a shark, the shark has no right not to be consumed as flake by humans.

The Case for Animal Rights (1984) contains 400 pages of carefully crafted argument that most animal supporters find daunting in comparison to Singer's more widely read and cited *Animal Liberation* (1975). In the preface to *The Case*, Regan explains that the book was written with three audiences in mind: supporters of the animal movement, his professional peers in philosophy, and finally those who work with animals, such as veterinarians and scientists. The book contains virtually nothing on the actual conditions of animals in the contexts that Regan seeks to abolish—in factory farms, in laboratories and in the wild—since its purpose is to provide a philosophical defence of animal rights in general and explain what it means to ascribe rights to animals. In the terminology used in *Compassionate Beasts*, Regan's book is about animal rights advocacy or claimsmaking, while Singer's *Animal Liberation* is directed more to the activist and gives detailed descriptions of the conditions under which animals suffer as well as prescriptions for action by those who want to support the cause.

Critics of the movement will exploit philosophical differences to discredit the movement as ideologically confused. Yet the differences matter little to the many supporters, advocates, and activists interviewed and surveyed for this book. Similarly, for those critical of the movement or for those who find the concepts of animal interests and animal rights questionable, the Singer/Regan position is essentially seen as *for* animal liberation and therefore untenable.

The Leahy/Frey Case Against Liberation

Pain

Leahy's intention of *putting animals in perspective* derives its inspiration from Wittgenstein's language games and theories about how language works. According to Wittgenstein, language works in the context of human experience, and a series of language games determines a word's meaning. Leahy's main idea is that animals are "primitive beings" that lack the power of speech and all that it entails, including desires, interests, and beliefs as we understand these concepts. Animals therefore cannot have beliefs because they lack language, and being speechless, they cannot have desires.

How is an animal protectionist to take seriously Wittgenstein's claim, as reported by Leahy, that to attribute pain to animals does not make *sense*? Pain in the Wittgensteinian sense applies to human beings and only in diluted form to lesser beings. This defies the experience of anyone who has had only long-term association with animals as companions, on farms, or in the wild. An official complaint of cruelty that animal activists lodged against an intensive piggery in Australia described "how two of the prolapses were seething with maggots and how *one pig screamed in pain* as another pig stood on its prolapse" (Barrowclough, Neales & Brew-Bevan 1997:16). Leahy's excessive adherence to the doctrine of language games would denigrate the activists' description as anthropomorphic, while the assertion that it makes no *sense* to describe the pig's scream as evidence of suffering is surely a failure of *common sense*, if not a sign of moral apathy.

The neo-Cartesian John S. Kennedy argues that it is dangerous to unthinkingly assume that animals are conscious (1992:157). To do so is to be guilty of the sin of anthropomorphism—the attribution of human mental experiences to nonhumans. Kennedy claims that it is not possible scientifically to determine from their behaviour whether or not animals have feelings.

Animal protectionists, like environmentalists confronted with the proposition that there is no evidence of an impending ecological catastrophe from global warming, would prudently invoke the precautionary principle to ensure that if there is any element of doubt about whether animals feel pain, then we should desist from treating them as insentient rather than unthinkingly assuming that they are unconscious. Kennedy seeks to dilute the severity of pain by referring to reports of people on the sportsfield or battlefield, or of those admitted to the hospital with severe injuries, who may be completely unaware of the pain.[1] In much the same way that environmental cornucopians serve vested economic interests, Kennedy's attack on the new anthropomorphism serves the interests of animal experimentalists.

Suffering

Leahy agrees with Singer that human beings *suffer* more than animals in many cases because animals, unlike humans, are incapable of pondering the associated torments—the loss of potential happiness, the possibility of living to a ripe old age, the opportunities and regrets about the past, and so on. Yet this has little to do with pain or cruelty as the focus of animal liberation campaigns. It is possible to argue, as some philosophers do, that pain is a greater evil than suffering.[2]

Many pet keepers armed with stories of animals pining for their lost companions, human or nonhuman, would contest the assertion that only humans suffer mental torment. And animal watchers, notably Masson and McCarthy (1995), would refute the proposition with their observation of the existence of complex emotions in animals, which Leahy would no doubt dismiss as an extreme form of anthropomorphism. Furthermore, it is not fanciful to suggest that animals in distress might suffer more than their human counterparts since, unlike humans, they are unable to engage in the capacity to exercise mind over matter to relieve their torment. Put more succinctly: "They have less capacity for distancing themselves" (Rodd 1992:73).

Leahy's perspective on animals is too abstract to be convincing to anyone other than those who seek to rationalise their exploitation of animals. Lawrence Johnson charges Leahy with "linguistic Cartesianism" and rightly suggests that the use of Wittgenstein's language games to counter animal liberation attacks against speciesism is as absurd as Descartes's "animals as machines" analogy of the seventeenth century (1993:5). It is true that humans alone are able to give meaning to mental cruelty or abuse by labelling such things as instances of torturing the soul. But this is not an argument that sentient beings incapable of articulating their pain or suffering do not experience it, as I have tried to show in the previous paragraph. Endless debates and language games about animal consciousness, or the nature of animal pain and suffering, will never satisfy individuals—like the Australian activists mentioned above—who have heard the pig's scream.

The American philosopher Raymond Frey is frequently invoked by opponents of animal liberation to support the moral orthodoxy of keeping animals in perspective. Yet Frey is as much a utilitarian as Singer; their philosophical differences are more of degree than of kind. Singer argues that most human uses of animals are unethical, especially those associated with factory farming, animal experimentation, and recreational hunting. According to Singer's calculus, the cost of these practices in terms of animals' lives and suffering greatly outweighs the benefits to humans, and therefore the practices are morally reprehensible. Frey's utilitarianism, on the other hand, sees the benefits of meat eating and improvements to human health, happiness, well-being, and so on as justifying the use of animals for these purposes. For him, animals can be eaten, dissected, and hunted provided that there is good reason to do so and that the killing is done humanely, that is in the absence of gratuitous cruelty. Frey rejects Regan's insistence that animals are subjects-of-a-life who have rights for all the usual reasons that critics of "rights talk" use to question the concept of animal rights. There is therefore more common ground between Frey's "concerned individual" and Singer's more pragmatic position on animal liberation than there is with the militantly abolitionist stance that Regan takes. What then do critics of animal rights have to say about the treatment of animals in commercial agriculture, in research, in zoos, and in hunting?

Eating Animals

The issue of meat is crucial to the animal protection cause since liberationists argue that if people believe that there is nothing wrong with eating animal flesh, then it is effectively a licence for humans to use animals as they please. Leahy's defence of killing animals for food seems to be that it is acceptable as long as there is no avoidable cruelty involved in their rearing, transport, and slaughter. The arguments in favour of vegetarianism—moral, health, environmental, and gustatory as canvassed by proselytes like Singer (1975), Robbins (1987), Fiddes

(1991), Cox (1992), and Hill (1996), among others—are given short shrift by Leahy. He suggests that even if we were to replace meat with grains, the benefits from the reduction of world hunger in underdeveloped countries would be outweighed by the chilling prospect of the "demographic and political implications" of a dietary revolution (1991:213), by which he no doubt means overpopulation and global instability.

Leahy rejects the idea that meat eating is a health hazard and claims that "it is only necessary to show that a diet to which [meat] contributes can be no less healthy than one from which it is excluded." At this point, Frey's 1983 book on moral vegetarianism is brought to the table, so to speak. Frey rejects Singer's claim that the costs to the meat industry as a consequence of a growth in vegetarianism are outweighed by the benefits to animals and goes on to list 21 instances that suggest that the opposite is true (1983:197–206). Leahy discusses just three of these issues—massive unemployment, forced dietary change and the demise of traditional cuisines and restaurants, and the conversion of a bucolic countryside into an animal-free zone—and contends that the Regans and Singers of the world who attempt to play down the consequences of a vegetarian future are insensitive to human welfare. Leahy and Frey exaggerate what Leahy calls "the possible catastrophic consequences of widespread vegetarianism for human beings" (1991:217) since all three of these consequences would diminish in severity within a generation or two. Intensive farming, on the other hand, appears to be becoming more intensive with ever increasing numbers of animals being killed in the process. It is difficult to see how, under such conditions, animal slaughter could become more humane or how any reasonable person would be convinced by Leahy's lame suggestion that thanatosis—a temporary paralysis or anaesthetic shock—could alleviate the suffering of factory-farmed animals as it evidently does for some accident victims.

Contrary to the anecdotal evidence of many animal watchers and people who work with animals, Leahy claims that animals, because they lack self-awareness, do not experience the uniquely human fear of death (1991:219). We have already seen that Leahy and other opponents of animal liberation maintain that we do not know if animals have feelings or experience pain as we understand it. Leahy's "linguistic Cartesianism" comes unstuck at several places, as when referring to ritual slaughter as "an agonising and unsavoury procedure lasting between one and two minutes" (1991:219). There seems no better reason to raise this issue other than as an attempt to sanitise, by way of comparison, the treatment and conditions of animals in British abattoirs. Yet on the manner of their slaughter and the fear of death that animals experience, Leahy is at odds with John Webster, a professor of animal husbandry. Webster contends that the difference between conventional and ritual slaughter is of minor importance, while the fear of death is the main problem facing animals in slaughterhouses (1994:106). Animals may not live in chronic fear of death as many humans do, but they do seem to know when their time is up. Under conditions of apparent cruelty, what can the humane individual do? A possible solution is offered by R. G. Frey, whose more practical arguments are oddly omitted from Leahy's account.

Frey's *Rights, Killing and Suffering* (1983) is subtitled *Moral Vegetarianism and Applied Ethics*. In this book, Frey promotes the idea that it is not necessary to be a vegetarian in order to do something about the plight of animals. The "concerned individual" is the quintessential animal welfarist—one who campaigns against the worst abuses of factory farming or speaks out against cruelty to animals without giving up meat or using animals in ways that do not harm them. Vegetarianism is not the only tactic available to anyone who is disturbed by our treatment of animals. Frey believes that a concerned individual should boycott meat produced cruelly but not meat that comes from a well-cared-for animal. He is against the more strict moral vegetarianism, which Singer and Regan advocate and is in favour of what might be called "semi-vegetarianism" (George 1994:405). For many people, these dietary regimes are preferable to the sacrifice or even oral masochism that is often associated with veganism and vegetarianism in predominantly meat-eating cultures. Semi-vegetarianism is popular also amongst animal activists (Munro 1996a) and is seen as a more realistic goal for women and others whose bodily needs are not the same as those of affluent, white males (George 1994). In proposing an alternative to moral vegetarianism, Frey offers a way out—purists would call it a convenient rationalisation—to those with a conscience who want to eat meat. His menu is therefore more palatable than what is offered by the militant school of vegetarians, vegans, and fruitarians whose practice of ethical eating seems beyond the capacity of most people.

Experimenting on Animals

Leahy's discussion of animal experimentation begins by noting both the uncertainties surrounding the practice within the science fraternity itself and the way that both sides present their respective moral and scientific claims to the general public. Wittgenstein's language games again reduce Leahy's arguments to little more than linguistic Cartesianism when he comes to discuss the concepts of pain, suffering, distress, and anxiety. Rowan's (1989:97) definition of animal suffering—"the unpleasant emotional response to more than minimal pain and distress"—is described by Leahy as "hopeless" because it ignores unwitting suffering (e.g., when plants suffer from too much sun) and is improperly anthropomorphic (when it covertly seeks our sympathy by referring to the animal's unpleasant emotional response to suffering). Who are we to believe on the subject of animal behaviour—the philosopher Leahy, who is opposed to animal liberation, or the biochemist Rowan, who has both experimented on animals and enjoyed good relations with the animal movement? Cautious readers might seek the views of an independent expert when confronted with doubts about such a choice.

Webster's credentials as a professor of animal husbandry put him in good stead to answer some of these questions. Webster seems to have no doubt that animals are capable both of suffering pain, distress and, anxiety and of experiencing emotional states. His professional judgement explicitly contradicts the line that Leahy and the neo-Cartesians take:

> Many animals, including those species whose lives we control the most demonstrate properties of mind which may be far less complex than ours but are sufficiently advanced to feel suffering and pleasure . . . Suffering occurs when the intensity or complexity of stresses exceeds or exhausts the capacity of the animal to cope, or when the animal is prevented from taking constructive action . . . Prolonged exposure to sources of suffering may induce abnormal behaviour patterns such as stereotypies or profound changes of mood, including the development of a sense of hopelessness. (1994:38)

Leahy may be wrong about animal behaviour, but his views are in accord with moral orthodoxy. He notes that rats and mice (among the least-loved animals) vastly outnumber all other species as laboratory animals, and he concedes that there is a strong case for banning the use of endangered chimpanzees (among the most charismatic of animals) in experiments despite the pressure of the AIDS lobby. The related issue of product testing puts rodents in a different light. Is it right to inflict pain and suffering on thousands of rats to test the safety of a new shoe polish or a more fragrant disinfectant? Leahy seems to put consumer interests before those of the animals when he expresses hostility to Regan's claim that consumers' interests should not be allowed to override the interests of animals to avoid pain and suffering. He does concede, however, that many pharmaceuticals, preservatives, insecticides, and so on are harmful to both humans and animals and that it would be in our mutual interest to reduce the number of toxic materials that come onto the market every week.

Hunting Wild Animals

The third area that Leahy covers—zoos, field sports, and furs—contains some of the remaining concerns of animal liberationists. Since zoos are not at the top of the list of grievances that concern animal protectionists, but hunting and trapping for furs are, I will focus on the latter, particularly on blood sports, which along with intensive farming and vivisection, represent the trio of campaigns described in this book. Leahy is at his most contemptuous when it comes to protests against hunting, shooting, and fishing. He claims that "much of the sound and fury seems to be generated more by a basic antipathy to the sorts of people who like to hunt and shoot and, in particular, by the fact that they do so for pleasure" (1991:244). While there may be a trace of class antagonism in the opposition to foxhunting in England, it is by no means the main reason for the opposition to blood sports there or elsewhere. Based on interviews with hunting opponents in the three case study countries, cruelty remains their main grievance.

In his discussion of our treatment of wild animals, Webster is at pains to point out that his primary interest is to ascertain the impact of hunting and shooting on the welfare of the animals, not to moralise on the rights and wrongs of our actions in either rescuing or killing animals in the wild. He is unequivocal in concluding that the chasing of deer and the hunting of foxes with hounds is cruel (1994:222).

Yet Leahy does not accept that the animal liberation movement's central grievance against hunting is the cruelty inflicted on wild animals. It is axiomatic that for anyone to be able to make a reliable statement about the motives of animal liberationists, the views of the activists themselves must be sought. As is very common on this and other issues discussed thus far, some philosophers tend to rely on abstract reasoning, or worse, pure speculation devoid of any empirical data. At the very least, Leahy ought to have acquainted himself with the independent assessment of animal behaviourists, not all of whom necessarily would come down on the side of the liberationists. Leahy's view that hunting is a form of pest control is, in fact, supported by Webster on the issue of fox hunting, although he thinks hunting them may no longer be necessary in these gentler times (Webster 1994:219).

Leahy claims further that traditional field sports are good for conservation and that many hunters are motivated toward "not only the conservation of the species they pursue, but the maintenance of the unspoiled countryside in which they have their natural habitats" (1991:247). For animal liberationists, however, cruelty inflicted against individual animals takes priority over conservation, which leads them into conflict with their otherwise kindred spirits in the environmental movement. Leahy is dismissive of the animal liberationists' cruelty talk since nature herself is red in tooth and claw. He suggests that it is futile to view nature as a state to be remedied and suggests that to condemn the hunters for inflicting "a certain amount of necessary pain in the otherwise justifiable cause of field sports" (1991:248) is akin to condemning nature for the suffering of animals in the predatory wild.

Animal liberationists would respond that humans are among the rarest of creatures, those who engage in recreational violence, and must be condemned for doing so. And unlike the wild animal that preys upon its fellows out of necessity, we are moral animals who have a choice. Leahy seems to think that hunters kill animals for several reasons, including pleasure, and that this is legitimate, especially if it is provided with a cloak of respectability on *other* grounds, such as conservation. The respected sociologist Alan Wolfe, whose thesis I discuss next, does not need any techniques of rationalisation to defend human exploitation of other animals.

Alan Wolfe and Human Exceptionalism

Alan Wolfe's *The Human Difference: Animals, Computers, and the Necessity of Social Science* (1993) takes a similar antiliberationist stance to Leahy's but from a distinctively sociological perspective. Wolfe is a proud speciesist who unblushingly manages to defend the worst excesses of anthropocentrism against what he sees as a dangerous antihumanistic cosmology in the form of social movements concerned with ecological issues and animal rights. Wolfe's concerns about animal rights are worth examining in some detail.

According to Wolfe, one essential difference remains between us and animals, and that is our capacity for interpretation and the production of meaning.[3] What makes us different, he suggests, is our power to use our minds to alter the rules that govern us. Unlike animals, we have the capacity to create meaning in our lives, which Wolfe acknowledges does not come without unfairness: "Just as we experiment on animals to keep humans alive, we are sometimes cruel to animals in order to give our lives meaning" (1993:91).

Wolfe's bluntness and unblushing speciesism would alarm not just animal liberationists, but ordinary fair-minded people. He goes on to explain how meaning-producing creatures need to exploit less-complex species to sustain their standard of living. As a consequence, there will be conflict between living by the principle of doing no harm and desiring to live "an imaginative and richly rewarding life" (1993:87). Animal experimentation is of central importance here since we sometimes use animals to keep us alive. And since we are nothing without life, animals genetically engineered to provide us with life-saving organs become our spare parts. Our ability to imaginatively recast the animal in this way to prolong our own lives is perhaps the most dramatic illustration to date of our total domination of animals. While Wolfe does not specifically condone this example of speciesism, he would not rule it out if it made our lives richer and more meaningful. As one of the movements engaged in *putting nature first*, he sees the animal rights movement as a threat to the good life, so defined:

> Animal rights theorists [he cites Singer and Regan] are thus correct to detect certain patterns of cruelty in the way we use other species to make our own lives more rich with meaning. At the same time, if we were to revise the ways in which humans make meaning out of the natural world in such a way as never to be cruel to other animal species, we would live in a world without *fantasy, excitement, and creativity*. (1993:87, emphasis added)

Fantasy

Wolfe singles out zoos and circuses as fulfilling symbolic roles in the lives of ordinary people. Zoos, he suggests, are not unlike poets. We can ban zoos in the way that Plato banned poets from the republic, but we would not necessarily be less cruel for doing so. Wolfe claims that children learn to develop their powers of imagination and fantasy by a visit to the zoo, a claim that is strenuously refuted by animal liberationists who suggest that what children learn about wild animals is "false and dangerous" (Jamieson 1985). Zoos tend to be seen as more benign than circuses since the circus animal has to be trained to perform unnatural tricks. Animal liberationists claim that this is evidence of extreme cruelty and oppose circuses for this reason. At the time of this writing, twenty-one local government councils throughout Australia have banned circuses as a result of lobbying from the RSPCA and other animal welfarists.

Wolfe would view these developments as unwelcome since they strip the child "of one of the many symbols of the interpretative life that children have always experienced, and the likelihood is that, as adults, they will be more unfeeling rather than less" (1993:88–9). A page later, he is dismissive of animal rights advocates for believing that if we treat other animals with more respect, we might treat our fellow humans more respectfully. According to Wolfe, the opposite may well be true, although he omits to say how and appears to contradict himself when he recycles an unproven cliche that "those who love animals have problems with humans" (1993:85). Wolfe could be thinking here of some of the prominent Nazis, but it would be unreasonable to include most animal liberationists, people who keep pets, or people who enjoy visiting animals in zoos and circuses in this rebuke.

Excitement

For many humans, especially those who live in big cities, the ultimate fantasy is to see animals in the wild. For many men and boys, however, the ultimate excitement is to hunt and kill wild animals. Were they to fantasise about killing their fellows, we would regard them as evil; on the other hand, the desire to hunt foxes or ducks or deer is viewed by many adults as deviant but by no means criminal. Thus, Wolfe is dismissive of Regan's (1983) observation that the oft-stated pleasures of hunting—exercise, communication with nature, camaraderie, and the satisfaction of the well-aimed shot—could be satisfactorily replaced by the more gentle pursuit of bushwalking with a friend whose only weapon is a camera. What hunters seek, says Wolfe, are "the sensual pleasures of violent sport" (1993:89), a claim ominously echoed in Jack Katz's 1988 book *Seductions of Crime: Moral and Sensual Attractions of Doing Evil.*

Creativity

Consider two ideas: (1) because we are human we are capable of creating our own worlds; and (2) human beings have always fantasised about living to a ripe old age without the inconvenience of illness or infirmity. Wolfe would no doubt approve of experimenting on animals if this was to help us combine these two ideas. Since animals are unable to think creatively and humans can, suggests Wolfe, we are therefore entitled to use nature for our own pleasure and well-being. Wolfe fears that the animal rights and environmental movements represent "one more nail in the coffin of anthropocentrism" (1993:11). He makes no pretense at disguising his hostility toward animal rights and the kind of society that the movement would spawn: "a society . . . poor in imagination, imprisoned by logic, censorial toward evil, puritanical in outlook, and populated by those as lacking in appreciation of themselves as they are sensitive to the pains of others" (1993:91). This is not a description of Nazi Germany or of anti-Semitism, but of animal liberationists using their minds to eliminate the worst excesses of speciesism and anthropocentrism! This anthropocentric sociologist's intemperate attack on animal rights clears the air for the accounts of animal rights activism and advocacy in the remaining chapters.

NOTES

1. Cicero once remarked that no opinion is so absurd that it has not been advocated by some philosopher. The same could be said for scientists, some of whom, like Kennedy, still claim that we cannot know whether animals have feelings or tell from their behaviour whether they suffer or not.

2. In the *The Decent Society*, the philosopher Avishai Margalit argues that such a society is one free of humiliation; people are subjected neither to humiliation by other people nor by institutions such as welfare agencies or prisons. Margalit believes, however, that humiliation runs a close second to the greatest evil—physical cruelty, especially the suffering that other human beings inflict: "Torturing the body causes more acute pain than torturing the soul" (1996:264). A decent society presupposes that physical cruelty has been eliminated. Although he suggests that cruelty toward man or beast is wrong, it is the suffering of humans not of animals, that interests Margalit. This is clearly implied in his explanation of humiliation as the treatment of humans as if they were animals, objects, or machines. While Margalit's *Decent Society* has little room for animals, he appears, unlike Leahy, to acknowledge that pain is pain whether inflicted on the body of a man or a beast.

3. Wolfe discusses nine claims about human uniqueness:

1. communication
2. metacommunication or talk about talk
3. the capacity for thought
4. symbolisation
5. toolmaking
6. culture
7. self-reflection
8. the division of labour
9. morality

The uniqueness to humans of some of these accomplishments has been challenged by scholars sympathetic to the animal rights cause who have drawn attention to human-animal continuities rather than to differences (Noske 1989; Haraway 1989).

PART II

GRASSROOTS ACTIVISM IN THE STREETS

Part II focuses on the world of the animal rights activist. I use the phrase "in the streets" to dramatise the grassroots character of animal rights activism in contrast to the more conventional lobbying of advocates "in the suites." In Chapter 4, I describe a hybrid form of animal activist and advocate who I believe is emblematic of, if not unique to, the Australian animal protection movement. The chapter describes the influence of Singer's *Animal Liberation* (1975) in the 1970s and early 1980s in Australia, particularly in the campaigns against factory farming. The role of women in the defence of sheep, pigs, and battery hens is also discussed. Chapter 5 begins with a brief discussion on the importance of compassion in the animal movement, particularly in the context of how people become and remain animal activists. How animal protectionists in Australia have managed to unite the different concerns of vegetarian, conservation, wildlife, and domestic animal groups into a pragmatic regional federation of animal societies is explained. Chapter 6 broadens the discussion of what it means to be an animal activist and describes the tactics that American, Australian, and English activists find most effective in their campaigns. A number of activist case histories are described, as is the relevance of ethical vegetarianism to the animal movement. In Chapter 7, which concludes Part II, I look at how collective action is organised to make the most of the mass media in campaigns to improve the lot of farm animals in the United States, the United Kingdom and Australia.

Chapter 4

The Prototype Activist-Advocate: An Australian Hybrid

> The philosophers have only interpreted the world in different ways; the point is to change it.
>
> —Karl Marx

Activism is defined in the Collins dictionary as a policy of taking direct action, often militant action, to achieve an end, especially a political or social one. Advocacy, according to Collins, refers to active support, especially of a cause, while an advocate is someone who pleads for or speaks in favour of a cause. As I argued in Chapter 2, different combinations of these two processes characterise movement structures in the United States, the United Kingdom, and Australia. In the United States, advocacy is more prominent than activism while the reverse is true in the United Kingdom. The Australian model is different again, with activism and advocacy being more balanced than in either the United States or in the United Kingdom. In this chapter, I look at what I believe is a unique Australian species of animal protectionist: a hybrid advocate/activist, the philosopher/activist Singer being the best-known example.

As I show in this chapter, Singer's willingness to take direct, nonviolent action on behalf of animals strengthens his advocacy against speciesism on philosophical grounds. His example—which stands as a corrective to Marx's dictum in the epigram to this chapter—has inspired many of the animal protectionists interviewed for this book and has led to the establishment of animal liberation groups throughout Australia and elsewhere.

In most cases, the SMOs in Australia were started by individual issue entrepreneurs who had the drive and commitment needed to build animal protection organisations from scratch. As it happened, with the exception of the Animal Liberation branches, the nascent SMOs had parent organisations either in the United Kingdom or in the United States on which they could be modelled, although the forms they took were quite different. The World League for the Protection of Animals (WLPA) in Sydney, Project Jonah in Melbourne, and the Sydney-based Australian Vegetarian Society are all member societies of the Australian and New Zealand Federation of Animal Societies (ANZFAS), the world's first regional federation of animal protectionists. In this chapter, I focus on Animal Liberation and in the next on ANZFAS and some of its member societies.

SINGER'S *ANIMAL LIBERATION* IN SYDNEY

In 1996 Animal Liberation groups celebrated their twentieth anniversary as animal protectionists in Australia. Outside the movement, there was little recognition of the anniversary and virtually no media coverage at all, apart from a singularly low-key radio program on the popular *Australia Talks Back*, hosted by the Australian Broadcasting Commission (ABC Radio 3 October 1996).[1] Singer, whose book *Animal Liberation* spawned the branch of the same name in Sydney twenty years earlier, was the program's special guest. The callers covered very little new ground, although a University of Queensland veterinary scientist claimed that pigs were "extremely aggressive animals" and were better off in pens than they were outside in the rain and mud. Singer, who has been arrested for attempting to liberate pigs in a New South Wales factory farm, replied that keeping sows in 1.8 by .6 metre pens and on concrete floors was cruel and against the welfare interests of the animal. It was appropriate that the plight of farm animals was the main topic of interest for the participants since farm animals were the motivating factor behind the formation of the Animal Liberation branch in New South Wales two decades earlier.

Riding on the Sheep's Back

The issue of live sheep exports was the animal movement's first campaign when Animal Liberation began in Sydney in 1976. The branch was the first animal protection group in the world to be based on the philosophy of Singer's *Animal Liberation* (1975). It was founded by Christine Townend after she read the book. From the outset, Animal Liberation (NSW) sought to project an image of respectability so as not to alienate ordinary Australians. Townend tells the story of the branch's origin and development in *A Voice for the Animals* (1981) and points out that it was the critical rural press and farmers' groups that gave the "animal libbers" the most publicity. Twenty years later, Animal Liberation branches exist in every state and are affiliated with the moderate, pragmatic Australian and New Zealand Federation of Animal Societies (ANZFAS), whose president is Singer. From the outset, farm animals have been the focus of the Animal Liberation branches, which constitute one of the largest membership sectors of ANZFAS (see Table 5.2 in the next chapter).

From 1981 until the present, ANZFAS has documented the extent of the live sheep trade to the Middle East and has argued that about 100,000 sheep die between ports each year. The SMO claims that the number is closer to 200,000 if the sheep who die in feedlots in the Middle East are included (ANZFAS Fact Sheet, August 1995:3). In the most recent disaster, a staggering 67,488 sheep perished at sea during a fire on board the *Uniceb* in September 1996. The minister for Primary Industries and Energy was not aware of the scale of deaths in the trade, evidently believing that the losses were "minuscule," according to ANZFAS director Glenys Oogjes. And the president of the Livestock Transporters Association of Western Australia, the state from which about three-quarters of the shipments originate, claimed losses in transit were less than one per cent. The RSPCA put the mortality rate at between 2 and 10 per cent and called for an immediate ban on the trade following the incident. This was unlikely to be taken seriously, given that the trade is worth $500 million a year to the rural sector in Australia. Furthermore, a senate inquiry in 1983 had noted that "if a decision were to be made on the future of the trade on purely animal welfare grounds, there is enough evidence to stop the trade" (Senate Select Committee on Animal Welfare 1985:xiii & 185). It is instructive to look in detail at this committee's report since it surveyed the views of the principal Australian interests in live animal exports.

In 1983 a new Labour government was elected in Australia, largely because of its pro-environmental platform. Although the environment was on balance the decisive issue, other factors were also important. Australia was in the grip of a long and harsh drought, and people were ready for a change of government. Anyone who had seen the Australian countryside during the prolonged drought could not have been unmoved by the sight of farm animals suffering under a scorching sun. Even in the national capital, Canberra, animals could be seen vainly searching for food in the dry paddocks not far from Parliament House. While these images had nothing to do with the Senate setting up the committee to inquire into animal welfare, it would be most unlikely that the suffering of animals that summer in Australia was lost on those attending the committee hearings. One would have expected that those representing the animal protection lobby to have at least received a sympathetic hearing.

Of the five issues to be addressed—animals in commerce, wildlife protection, animal experimentation, codes of practise of animal husbandry, and animals in sport—the live sheep trade was the second matter to be examined after kangaroo hunting, the scale and savagery of which had attracted public concern both in Australia and overseas (Munro 1993/1994:17). Although the scale of the live sheep trade could not match the quota of nearly five million kangaroos "culled" legally in Australia, it does represent "the largest, planned, mass movement of animals by sea in the history of the world" (Senate Select Committee 1985:9).

In its report of over 200 pages, the committee made approximately 30 recommendations after hearing evidence and receiving submissions from dozens of expert witnesses and interested parties.

When the trade in live sheep began in the 1830s, there was a small fleet of no more than 20 ships, each carrying 300 to 1,000 animals; that is, probably no more than 15,000 sheep would be exported yearly. A century later, when "Australia rides on the sheep's back" became a national motto, the numbers had risen to 6.25 million in 1980 and to nearly 7 million in 1984 (Senate Select Committee 1985:101). It is the sheer scale of the trade that has led to conflict between animal welfarists and the economic interests involved. For example, the first line of the Australian Veterinary Association's (AVA) policy is to acknowledge that the trade is necessary for the industry's economic viability (1993:81). Human interests, including the employment of qualified veterinarians to inspect the animals, are more prominent in the policy statement than animal welfare considerations are.

One chapter of the Senate's 200-page report, "Animal Welfare and Economic Considerations," devoted just over six pages to the cruelty versus commerce argument that concerns animal protectionists everywhere. The chapter refers to various philosophical arguments about facts and values, in particular Regan's dissatisfaction with Singer's utilitarianism, and concludes that the consequences for all those involved in the trade would not necessarily be better if it were banned. In other words, Singer's cost/benefit calculations would not be able to weigh the interests of the animals in any meaningful way since, from a moral standpoint, the value of an animal's life cannot be quantified. Regan's hard-line animal rights approach is usually cited to ensure that human rights prevail over those of other animals, since "rights talk" is an invitation to an endless debate between philosophers and laypersons alike. Thus, the Senate Select Committee on Animal Welfare in 1989 was told that "rights talk" creates the problem of a hierarchy of rights of various species: "Would there be specifically animal rights, tiger rights, pelican rights, tape-worm rights?" (Senate Select Committee Report 1989:35).

On the other hand, the committee noted that "rights talk" has become popular because it gives the bearer moral, if not legal, protection. The committee concluded that each case involving conflict between the rights of humans and animals must be decided on its merits, and not necessarily in favour of humans (Senate Select Committee Report 1989:42). These pious sentiments need to be considered in light of what actually happens when human interests are pitted against those of animals. The case of the Senate inquiry into the kangaroo industry, for example, recommended that the kangaroo remain a protected animal but accepted that they could be killed to protect property if the level of damage was unjustifiable and there were no alternatives to slaughter (Senate Select Committee Report 1988:176). Many animal liberationists believe that government inquiries into animal welfare are inauthentic because they invariably privilege human welfare and the interests of the animal-user industry.

WOMEN AND ANIMAL ACTIVISM

A perusal of the list of witnesses to the senate inquiries on animal welfare in Australia reveals an industry bias that reflects male occupational interests in fields such as intensive livestock production, commercial use of wildlife, animal experimentation, and horse racing. And the Senate Select Committee itself tended to be dominated by male senators despite the trend during the 1980s and early 1990s of an increasing number of women being elected to the Australian Senate. Animal welfare interests, on the other hand, were and still are dominated by women, a characteristic of the animal movement worldwide.

Soon after Animal Liberation was established in Sydney, more branches were established throughout Australia so that by the early 1980s, most states and capital cities were represented.[2] Women established all but one of the seven branches which raises the intriguing question as to the massive gender imbalance in the animal movement as a whole since women outnumber men by three or four to one (Jasper and Nelkin 1992; Richards 1992). A starting point in explaining the gender imbalance can be made by referring to some of the many scholarly books that deal with the theme of animals and women. For example, Adams and Donovan, as editors of *Animals and Women* (1995), include a bibliography of about 150 works on this theme and on the topic of feminism and animal defence theory. To my knowledge, there is no comparable literature on men and animals, no doubt because it is women, not men, who share with animals the status of an oppressed group. Rheya Linden, the president of Animal Liberation Victoria, points to the common interests of women and animals:

> *I think we need to look at the politics of animal rights and to see how they do converge quite clearly with feminist politics. I think it's very easy for women who've been through the feminist movement particularly, and have matured through that, and their daughters to see the issues in animal rights. And they are issues of oppression. They're issues of abuse, and the link is very easy to make for women because women have known what it is to live in patriarchies, to know what it is to confront that masculine scientific detachment that allows abuse to continue for abstract, greater goals.* (Interview August 1997)

Jasper and Nelkin give a number of more conventional reasons for women's massive overrepresentation in the animal rights movement, such as greater availability of time and the persistence of traditional gender expectations that equate expressive, rather than instrumental roles, with women (1992:39). They attribute the long association of women in the early animal protection movements of the nineteenth century to the opportunities that animals provided for caring and compassion, qualities that were socially acceptable for women to cultivate in humane work (1992:59). And as noted in Chapter 1, in the eighteenth and nineteenth centuries, a number of prominent women made the connection in their campaigns between the exploitation of animals and the abuse of women.

That women empathise with animals more than men do is understandable given the fact that men dominate in virtually every major animal-oriented occupation and leisure activity—in hunting and trapping, in abattoirs, in research laboratories, in veterinary medicine, on intensive farms, in sports such as shooting and angling, and in a host of other pursuits where humans use and abuse animals. It is therefore not surprising to find that of the twenty-seven animal protection organisations represented in this study, women head slightly over half, although only three of these are large, high-profile SMOs; in contrast, half of the organisations that men head are in this category. Men are typically employed in well-paid, senior positions in advocacy organisations, while women generally work as volunteers in grassroots or smaller advocacy SMOs. It is interesting to compare these gendered work patterns with the opposition in antienvironmental organisations such as Put People First (PFF). A sample of the same number of these organisations, listed in a Greepeace Guide (Deal 1993), revealed that of the twenty-seven selected organisations, men headed twenty and women headed only seven. Again, this can be explained by the industries and interests that these antigreen organisations represent. Most are male-oriented enterprises associated with the extractive industries (coal, gas, oil, and timber), off-road vehicle manufacturers, hunting and fishing lobby groups, chemical and pharmaceutical companies, the cattle industry, and so on. There can be little doubt that leadership positions in both the animal movement and the antienvironmental movement reflect the structure of gendered employment opportunities in mainstream society.

However, in at least one activity there is equality between the sexes, namely meat eating, which Adams describes as "the most oppressive and extensive and institutionalised violence against animals" (1990:70). It is this issue that concerns many of the feminist writers and activists in the animal movement. For example, in the opening sentence of an article on speciesism and sexism, Emma Munro makes the provocative claim that "on a global scale the most exploited humans are women and in factory farming the most exploited animals are female" (1997:56). Adams has also drawn attention to the links between speciesism and sexism; in one section where she highlights the connection, she specifically focusses on the treatment of pigs. Karen Davis (1995) has done the same for chickens, while others have demonstrated how our language reflects sexist undertones. According to Nilsen (1977), the animal comparison spans a woman's life: a young girl is a *chick*, leaves the *coop*, goes to *hen parties* and enjoys the *cackles* of her friends, but has to return to her *brood*, where she *henpecks* her husband, and ends up an *old biddy* (cited in Dunayer 1995:12). Dunayer interprets these terms as metaphors of exploitation: the chick is the young female designed for giving sexual pleasure while the biddy has outlived her sexual and reproductive usefulness: "If hens were not held captive and treated as nothing more than their bodies, their lives would not supply symbols for the lives of stifled and physically exploited women" (1995:12–13). Davis, who founded United Poultry Concerns in 1990, makes many of the same points about caring for chickens that Patty Mark, the founder of Animal Liberation Victoria makes about the animals she has rescued from battery hen installations. Davis writes warmly of Viva, "the first chicken I ever really knew," and of extending her "acquaintance" to turkeys, ducks, and geese (1995). Patty Mark also believes that it is necessary for activists to "know" animals if they are to promote their welfare:

I've studied hens quite a lot, and I've kept them now for six or seven years, which is the best thing to do. If you really want to know an animal, live with them! Even though it's been so many years, when I think back to the early 1980s, how I talked so much for battery hens, but never really knew them, how more important it is now that I know them inside and out, like people who keep a dog know the dog. I think it's so important that you really know your animals. Then you can really speak, not only just from your head, but from your heart as well and from experience. (Interview 1994)

The point in referring to the sentiments expressed by these female activists and founders of animal rights organisations is to suggest that these ideas would rarely be put in such heartfelt terms by their male colleagues in the movement. At least this has been my experience from the interviews conducted for this book. On numerous occasions, women, rather than men, spoke of their "bonding" with animals as if to imply a sense of shared suffering and an experience of oppression that was unique to animals and women. For Rheya Linden, *"caring about blood, flesh and pain"* is what makes animal rights important to women and what enables them, more so than men, to empathise with the suffering of animals:

And to turn it around again and put it right flat bang in a feminist perspective, I'd have to say that people who care about animals, and are prepared to politicise that caring, care about blood, flesh, pain, care about a particular animal's suffering in this particular situation now. So they are situating their caring, they're not abstracting it. (Interview 1996)

LIBERATING PIGS AND BATTERY HENS

More than any other animals, with the possible exception of animals used in research, the plight of battery hens and pigs in factory farming illustrates most dramatically the suffering that comes from the oppression of one species by another and the passion that is involved in their liberation.

Some five million pigs, 15 million lambs, seven million beef cattle, one million veal calves, and 345 million chickens are consumed in Australia every year. Of the 210 million dozen eggs produced each year, 95 per cent are derived from caged birds (Barrowclough, Neales & Brew-Bevan 1997:22–3). These figures, although only a fraction of the number of farm animals slaughtered in the United Kingdom and the United States, are a stark reminder of the massive number of animals' lives that are destined for the kitchen every day in meat-eating societies like Australia. Although the consumption of meat and meat products in Australia has been on the decline since the end of World War II, pig and chicken meat are an exception. In the 1950s, Australians consumed chickens only on special occasions such as Christmas. By the late 1990s, the number of chickens destined for the dinner table had increased more than a hundredfold (Barrowclough, Neales & Brew-Bevan 1997:23).

Some idea of the morality of the intensive animal industry can be gauged from a standard textbook on agriculture in Australia, in which the matter-of-fact description suggests the pervasive factory-farm mentality that animal liberationists oppose. Farmers unblushingly use the term "intensive farming" to describe the inhumane conditions that animal activists denounce. The textbook points out that "the aim of the egg producer is to convert a given amount of feed into as many eggs as possible" and notes that "top managers with top hens achieve over 280 eggs per hen" (Malcolm, Sale & Egan 1996:168). According to the Vegetarian Society of the United Kingdom, the production from hens in the wild is in the region of 12 to 20 eggs annually (VSUK information sheet 1996). Where the Vegetarian Society focuses on the cruel conditions that the animals endure in intensive farming and the health consequences for the birds, the agriculture textbook is only concerned with the economics of battery egg production, explaining for example, how more eggs can be produced by forced moulting, which can be achieved by reducing the natural light available to the birds to eight hours per day (Malcolm, Sale & Egan 1996:169).

The use of animals in research and for the mass production of food have preoccupied most of the Animal Liberation branches in Australia for the past two decades. These issues were also the subject of inquiries by the federal government in the mid-1980s. The Senate Select Committee on Animal Welfare in 1990 recommended that the tethering of sows be banned and that sows be housed away from individual stalls; a further recommendation, curious in its reasoning, drew attention to the increasing body weight of sows and the need to ensure slightly larger farrowing crates and stalls so as to prevent cramping.

Even more curious was the submission from the Australian Veterinary Association that stated that confinement systems of housing pigs were likely to be more conducive to good pig welfare than rearing in outside systems would be (Senate Select Committee 1990:285). According to Webster, covered yards and outdoor arks are considerably more humane than sow stalls (1994:148–9). He points out that for outdoor sows, the risk of pain and injury (other than sunburn) is very low, and aggressive behaviour among the animals is seldom a problem. The AVA's submission also included confirmation that the codes of practice developed for the welfare of pigs satisfied the association's standards of pig husbandry (AVA 1990:285). While this may be the case, ANZFAS director Glenys Oogjes points out that the Australian Codes of Practice all include statements supporting the "five freedoms," which mandates freedom from the following:

1. hunger and thirst
2. discomfort
3. pain, injury, or disease
4. fear and stress
5. freedom to express normal behaviour

However, according to Oogjes, there is considerable difference in interpretation among government, industry, and animal welfarists. The range of possible interpretations becomes even greater when one adds different interests within these groups, such as government and private veterinarians, farmers, transport companies, stock agents, saleyard personnel, the RSPCA, and other animal welfare organisations.

It is unlikely that people other than these vested interests have much conception of the way that pigs are housed in Australia. Animal protectionists have been campaigning for years against the cruelty to pigs confined and tethered in factory farms throughout Australia, where five million pigs are annually slaughtered; more than 95 per cent of the pigs are confined in intensive conditions for the duration of their lives (Oogjes 1997:14). The plight of pigs has not been an easy one for the animal welfare lobby to dramatise since like the battery hen, the pig does not have the "cute-and-cuddly" appeal of the native koala or domestic cat and dog. It was only when it was revealed that the then Australian prime minister, Paul Keating, owned part of a large piggery in Scone, New South Wales that animal protectionists were provided with an opportunity to bring the welfare of pigs to the attention of the public in the early 1990s.

The Case of the Prime Minister's Pigs

Activists from the New South Wales, Australian Capital Territory and Victorian branches of Animal Liberation targetted "Paul's piggery" and challenged the prime minister to improve the conditions at the factory farm. The activists used a number of tactics in the campaign, ranging from persuasion to peer pressure, to get the prime minister to show some leadership in the humane treatment of animals. Various celebrities and VIPs encouraged the national leader along this road. Paul and Linda McCartney wrote to the Prime Minister, telling him that "you're not kind and fair if you're murdering animals for profit" (1993a:8). Singer also wrote to Keating about the living conditions of the 25,000 pigs but received only a cursory reply from the Prime Minister's staff. Helga Kuhse, a philosopher and colleague of Singer's, wrote a feature article about the rights of animals that included Singer's comment that "the condition of [the PM's] pigs is properly a national issue," adding that pigs "are the most intelligent of all the animals consumed in the Western world" and therefore have different needs from less intelligent ones (Kuhse 1992:13).

It was not until animal activists, including Singer and Australian actress Lynda Stoner, attracted media attention with a stunt at the piggery that the issue came to the public's attention. About 80 activists—mainly from Animal Liberation branches throughout eastern Australia—converged on "Paul's piggery," seeking permission for a delegation to inspect the farm. When this was denied, Singer, his young daughter, Esther, and Stoner entered the farm to get photographic evidence of the sows' conditions. Singer and Stoner were arrested soon after and got the publicity that the activists were seeking, including a front page picture of the photogenic Stoner in the *Sun Herald* 29 November 1992.

Singer himself believes that the media coverage of the stunt was widespread in Australia and suggests that Keating's prominence as the part-time owner of the pig farm has made it possible for activists to educate millions of Australians about the way in which pigs are kept. While this may be the case, the timing of the publicity was unfortunate, as most Australians would be focussing at year's end on the summer break and Christmas. And for most people, this means the consumption of factory-farmed animals, including pigs. However, the story of the national leader's interest in pigs did attract some unwelcome publicity for a prime minister facing what most commentators wrongly predicted would be his last election.

The activists continued to exploit the prime-ministerial interest in pigs as a means of attracting media coverage for their anticruelty campaign. The *Animal Liberation Magazine* ran a number of articles about "Paul Keating making money from pig suffering" (*Action* 1993a:45, 8), "Keating's piggery to quadruple in size" (*Action* 1993a:46, 17), and described the "filthy, barren hellhole" where the tethered sows were kept (*Action* 1994:50, 8). The Animal Liberation activists used graphic photographs and video footage to support their claims that the prime minister's pigs were being treated cruelly. The activists called on members to attend a protest at Parliament House in Canberra in late 1993. During this protest, some activists heckled the prime minister from the public gallery and were promptly removed by security staff. But the biggest story occurred when the piggery sued Stoner for defamation in relation to a magazine article on the demonstration at which she and Singer had been arrested. A source inside the movement subsequently told me that the issue was settled out of court.

Unlike cases where the activists have little to lose by going to court, SMOs like Animal Liberation are reluctant to get involved in expensive legal proceedings that can ruin them financially. Instead, Animal Liberation branches in Australia have pursued a course of legal, direct action and conventional lobbying. Many of the campaigns have been educational, print-based ones that were not designed to stir the imagination of the public or to capture the attention of the media. This at least has been the public face of an organization anxious to convey an image of conventionality, if not respectability. To its members, however, a radically different identity is constructed. For instance, *Animal Action*, the official magazine of the organization, reveals a much more militant and angry constituency who risk jail sentences, rescue animals in dangerous places, eat "nonviolent" food, expose animal welfare bureaucrats who are soft on animal exploiters, and reveal the consequences of inaction with graphic images of suffering animals. Animal Liberation thus portrays itself as having two souls, the tame and the wild, so to speak. Its dual identity is emphasised in *Animal Action*'s respectable cover, which is starkly at odds with the atrocity stories and horrific images inside the magazine.

Yet animal liberationists in Australia are moderate in their activism. Clandestine tactics, such as the investigative surveillance of factory farms, are the most extreme measures that they likely to take. Late in 1995, for example, 32 activists tethered themselves alongside the sows at the Scone piggery, an act that for the first time made a difference to the welfare of factory-farmed animals. Immediately after the police charged the activists with trespassing, the New South Wales minister for agriculture announced a ban on animal tethering, starting in 1996. According to the president of Animal Liberation (NSW), Mark Pearson, this was the first significant victory for intensively farmed animals in Australia (Pearson 1996:18). That it was only achieved by activists prepared to be arrested angers Pearson, who believes that any law that condones cruelty for commercial profit is immoral. This is also the basis for the protests against live exports in England, outlined in Chapter 7. There, as in Australia, the protesters used a similar sit-in tactic to demonstrate their empathy for the suffering of the farm animals. Margaret Setter, one of the staff at Animal Liberation (NSW), wrote movingly of the "shared experience" of pigs and people "each one chained by the neck . . . opposite a row of tethered sows. We were greeted as friends . . . snouts were caressed, backs scratched—while the animals munched on unaccustomed treats of fruit and salad vegetables" (*Animals Today* February-April 1996:19).

Speaking Meat

By a quirk of fate, pigs became the filmic flavour of the year in 1995 because of the Australian-Hollywood production of *Babe*, which captivated millions of moviegoers throughout the English-speaking world. Animal protectionists were quick to realise the film's potential for mobilising support for the cause of factory-farmed animals. The attractive and intelligent animals depicted in *Babe* had the effect of causing many people, at least temporarily, to swear off bacon and pork, the demand for which fell, much to the consternation of producers. The animal movement achieved a brief moment in the sun when the actor who played farmer Hoggett in *Babe*, James Cromwell, "came out" at the March for the Animals in Washington D.C. and declared himself a vegan convert. He evidently attributed the conversion to the experience of making the film.

Others have commented on the film's power to challenge our assumptions about what animals are: "As his mother is prodded into the truck, Babe utters his grief so fleetingly and naturally that we hardly notice that our usual assumptions have been turned on their heads. The meat animal is being presented to us as an expressive, narrative subject—the meat is speaking" (Plumwood 1997:25). According to Plumwood, the idea of "speaking meat" thoroughly disrupts the Cartesian stereotype of the machine-animals that spend their entire lives in factory farms for the sole purpose of serving the human appetite. We are meant to see in *Babe* an example of the living, communicating, sentient being whose ontological presence is cruelly and ruthlessly denied in the factory farm. While cynics might complain that this is sentimental anthropomorphism in the extreme, there can be no doubt that many of those who saw the film, for a brief moment at least, perceived pigs in a wholly different way. And for animal protectionists, it is this transforming moment that they seek to exploit with promotional T-shirts, campaign brochures, slogans such as "Keep pork off your fork," and the like.

Yet the cruelty toward pigs in Australia continues. At the Bunge piggery in Corowa New South Wales 230,000 pigs are kept in the largest factory farm for pigs in the Southern Hemisphere. Animal Liberation (NSW) mounted an undercover rescue of the animals on Remembrance Day 1996. They described the facility as the "pig city of despair." In language usually reserved for describing crimes of the magnitude of genocide and gang rape, the president of Animal Liberation (NSW) reluctantly wrote of "an Auschwitz, a Gulag and an Alcatraz" with references to screaming, abused sows awaiting the torture of "the rape shed" (Pearson 1997:6–7). For people outside the movement, or for anyone who has not witnessed a slaughterhouse or factory farm of this size, the description may well seem extreme. However, activists like Pearson have, for many years, fought against official indifference and laxity in enforcing the Codes of Practice, which ostensibly protect farm animals from cruel and unnatural treatment. It is not surprising that they resort to extreme language, if not tactics, to express the depth of their concern for the animals.

The most recent manifestation of the activists' frustration over the treatment of intensively farmed animals was their expose on national television of the plight of pigs and battery hens. Although it is rare that animal welfare stories feature in the television news in Australia, the story of Animal Liberation's allegations against the RSPCA in New South Wales was one that the media could not ignore. The story ran for just under ten minutes on the *7.30 Report* (ABC TV 7 April 1997), a respected national current-affairs program that attracts a large audience throughout Australia. At the outset of the broadcast, the compere warned the audience that some of the footage taken by the activists might be distressing to some viewers; he remarked at the end of the story that the ABC had refrained from showing the worst examples of cruelty out of deference to viewer sensibilities. Footage taken undercover by animal liberationists at the Bunge piggery in New South Wales and at the Parkwood Eggs battery hen farm in the Australian Capital Territory showed sick and grossly injured animals, of the kind revealed to readers of the *Action* magazine.

To get media attention, activists are compelled to rescue the animals and to engage in covert investigations because, they claim, "the RSPCA and the Department of Agriculture stand united in refusing to take legal action" (*Action* 1995:53, 10). In the aforementioned television report, the RSPCA president acknowledged that animal liberationists keep his organisation honest by taking direct action on behalf of animals. It would appear that the RSPCA itself is unlikely to take action against the large-scale, systematic exploitation of animals in the mass production of food. Evidently, intensively farmed pigs and battery hens are too numerous and the institutionalised cruelty too entrenched for the RSPCA to intervene. In New South Wales for example, there has only been one prosecution against intensive farms between 1993 and 1997 (Barraclough, Neales & Brew-Bevan 1997:16). In their article that raises the question of the RSPCA's privileging of cats and dogs over farm animals, Barraclough et al. report an RSPCA official who acknowledges that the society has been discouraged from pursuing cruelty cases because appeals were costing it around $500,000 a year.

Pearson believed that there was something sinister in the RSPCA's failure to protect hens and pigs from the horrific conditions in factory farms in New South Wales and the Australian Capital Territory. Animal Liberation believed that the RSPCA (NSW) was doing a deal with the farm lobby, an allegation that the RSPCA strenuously denied. The president of the RSPCA Australia, the veterinarian Hugh Wirth, agreed that factory farming of the type shown on the *7.30 Report* was "institutionalised cruelty" but claimed that the RSPCA was powerless to act because they were "snookered by the law" (ABCTV 1997). And as one of the leading egg producers remarked during the program, under the terms of the Animal Welfare Act of New South Wales (1979), factory-farm animals were not treated cruelly. Much depends on how the animal welfare codes and laws are interpreted. Some explanation of legislation for the protection of animals in Australia is necessary if we are to understand what animal protectionists are up against when they seek legal means to prevent cruelty to animals.

The Right to Own a Beak

A recent cover story in *The Economist* featured a battery hen on the cover under the heading "What we owe to animals." The editorial opened with a description of the bird:

> She is confined to a tiny cage with four or five others for her entire adult life . . . squeezed into a space about the size of the picture on our cover, barely enough to move. She may exercise her pecking instinct by pecking out her neighbours' feathers, unless her beak has been cut off with a red-hot blade, probably causing pain for life. (*The Economist* 19 August 1995c: 11)

The question that animal liberationists raised in Britain and elsewhere boils down to whether or not a hen has a right to own her beak. In *The Economist* editorial, the issue of animal rights is discussed as a noble but futile exercise, since "without agreement on the rights of people, arguing about the rights of animals is fruitless." Elsewhere in the feature story, the prevailing orthodoxy is described as being able to do what you want with animals you own, save when the law says otherwise (*The Economist* 1995c:18), which applies equally well in Australia where animals are defined as property. Animal welfare leglislation in Australia is essentially moral legislation designed to "protect the disadvantaged from the excesses of the powerful" (Blackman et al. 1989). Prevention of cruelty to animals acts in the various states and territories are concerned with "aggravated cruelty," "neglect," and "lack of action" under criminal law. But contrary to the intention of the anticruelty acts, the Codes of Practice favour the animals' owners, not the animals themselves.

While the individual states and territories are responsible for animal welfare matters, the national government has drawn up the Australian Model Codes of Practice for the Welfare of Animals, which are meant to guide the states and territories. The Codes of Practice focus on welfare issues that might be associated with systems and procedures of animal husbandry; but because the codes are not binding on the states and territories, there is much confusion and inconsistency on how the codes are interpreted. This is particularly obvious in the case of layer hen welfare, which has been the subject of a call for national management. Gregory (1994), for instance, argues that legislative force is needed to coordinate an Australia-wide phase out of battery farming because the Codes of Practice are proven to favour commerce over cruelty. He argues that the codes serve the interests of capital since the animals are defined as property, rather than as sentient beings deserving of moral consideration. Under these conditions, even gross breaches of cruelty to animals are not allowed to infringe on the private property interests of individual farmers. But as Gregory points out, the cruelty of intensive systems is unnecessary as battery facilities have been phased out elsewhere without serious consequences for either consumers or farmers.

Let the Hen Walk

The plight of battery hens has been the main concern of the founder of Animal Liberation (Victoria), Patty Mark, who is also the editor of *Action* magazine. Virtually every edition of the magazine has covered the issue with reports of activism in the various states and territories, replete with the gruesome pictures procured during rescue operations. Like most of the animal rescuers described elsewhere in this book, activists like Mark seek to liberate animals by peaceful means. Mark is optimistic that the public will respond compassionately to the plight of battery hens once they see the cruelty involved in egg production. She claims that people would be prepared to pay more for the eggs to *let the hen walk*. Animal Liberation's pictures and video images frame the battery hen campaign as a rescue operation, rather than as the kind of angry protest that the media like to report. Mark says that she finds direct action "nerve-racking" but is "frustrated and angry" by official indifference to animal suffering:

> *What do you do? . . . You've gone to the police, you've gone to the minister, you've gone to the RSPCA, you've done everything legally viable, and nobody does anything. Then I think I have a moral responsibility to individually go in and help those animals. And so I'll be straight, at the same time, I don't want to do that, I don't like to do that, it's nerve-racking.* (Interview 1994)

There are signs that the cruelty/commerce nexus is being questioned in some Australian jurisdictions. In one recent case in Tasmania, a magistrate found that battery hens had been treated cruelly despite the treatment being in accordance with the Codes of Practice. In the Australian Capital Territory recently, the magistrate presiding in the Parkwood Eggs case referred to "the savagery of the system" of battery farms that, he believed, justified the "mild trespass" by four animal liberationists to rescue the hens (Ward 1997:11). The inadequacy of the code in relation to the Australian Capital Territory is apparent in the wording of the code, which would allow the proprietor of a battery hen farm to deny liability for causing suffering or pain to the hens under his control. In fact, this was the view of the magistrate in the Australian Capital Territory case. In most court cases involving animal rights activists in Australia, the magistrates have tended to favour the proprietors whose animals and/or "property" was the subject of the case. However, in the Australian Capital Territory case, the magistrate was of the opinion that the activists did not commit an offence in interceding to rescue the hens and noted: No doubt the code of conduct is a compromise between animal welfare and making a profit but as magistrate Ward pointed out, "the purpose of the Act is to promote animal welfare not destroy it" (Ward 1997:2).

The magistrate went on to say that despite a government animal welfare agency giving the "all clear" to Parkwood Eggs, he agreed with the animal liberationists that producing eggs by battery farming was "inherently cruel" and noted that the evidence for this came from the activists as well as from the prosecution expert witnesses. A defence witness gave six reasons why the battery system failed to consider the behavioural needs of the birds, including the freedom to move and the freedom to own a beak, all of which the magistrate noted are demanded by the Codes of Practice. This prompted him to inquire, "[h]ow can the code demand, on the one hand, as a basic requirement for the welfare of poultry, a system appropriate to the hen's physiological and behavioural needs, and on the other hand, allow for a system inimical to those needs?" The activists had also repeatedly asked the same question of the animal welfare authorities, only to be met with a brick wall of indifference. In the Parkwood Eggs case, the magistrate drew attention to the reasonableness of the defendants' actions, given their frustration over the failure of officials to acknowledge their legitimate concerns.

The Australian Capital Territory case suggests that the tide of public opinion, as indicated by the magistrate's ruling, may be changing in a direction favourable to animal liberationists who have been fighting a losing battle in the courts, particularly concerning the welfare of battery hens. Animal activists elsewhere who are engaged in campaigns to ban battery cages will no doubt take heart in this example of what Karl Weick calls "the psychology of small wins." He defines a small win as "a concrete, completed, implemented outcome of moderate importance. By itself, one small win may seem unimportant. A series of wins at small but significant tasks, however, reveals a pattern that may attract allies, deter opponents, and lower resistance to subsequent proposals" (1984: 43). The psychology of small wins[3] is also what drives the animal protectionists in ANZFAS, whose activities are described in the next chapter.

NOTES

1. Media coverage of animal welfare issues in Australia is generally slight. Apart from television coverage of the annual duck-shooting season, the issue rarely attracts sustained coverage in the press. An exception occured in the early 1980s when a series of articles in *The Age* focussed on the pros and cons of animal welfare issues. This was a deviation from the usual media story of conflict between activists and farmers. The issues covered were vivisection and factory farming, the two examples of speciesism that Singer chose to highlight in *Animal Liberation* (1975).

2. The various state Animal Liberation branches in Australia were founded in the year indicated by the following individuals:

New South Wales, December 1976, Christine Townend

Victoria, December 1978, Patty Mark

Queensland, December 1979, Jacqui Batzloff

The Australian Capital Territory, October 1980, Jenny McDougall

Western Australia, August 1981, Ilse Howard

South Australia, 1981, George Karoli

Tasmania, October 1982, Pam Clarke

3. In 1999, the Australian Capital Territory government, acting on the initiative of the Green Party, introduced legislation to ban battery cages in the territory. In 1992, Switzerland became the first country in the world to outlaw battery cages. Early in 1999, the European Parliament decided to do likewise, starting on January 1, 2006, an event that Compassion in World Farming described as "a tremendous day for animal welfare" (Alcorn, 1999:11). In the case of the Australian Capital Territory, the victory is more modest in that there is no obligation for the remaining six Australian states and one territory to follow suit. Even so, as only the second jurisdiction in the world to ban battery cages, the victory is a good example of the psychology of small wins since it means that at least one Australian jurisdiction acknowledges the cruelty of producing eggs in battery facilities. Moreover, the decision vindicates the claims of animal protectionists everywhere that hens have a right to be spared unnecessary suffering.

Chapter 5

Animal Rights Amateurism and Professionalism

Professionals keep the movement organized. Amateurs keep it honest.

—Stephen Fox

Stephen Fox's succinct comment neatly sums up how many animal movement insiders and analysts distinguish between the professionals and amateurs in the movement. Yet what unites animal protectionists in the suites and in the streets is a shared sense of compassion for other animals. Emotion, as Groves (1995, 1997) has argued, is an important but neglected component of social movement activism. Moral resources or moral capital—in the form of people's compassion for animals—is fundamental to the animal liberation cause, based as it is on caring for a species that is not our own. For Finsen and Finsen (1994), compassion is the moral foundation of the animal movement.

COMPASSION: THE EMOTIONAL HEART OF THE MOVEMENT

Joan Court, an animal rights activist in Cambridge, believes that pity and anger—"*and a general sort of compassion*"—is what motivates most animal protectionists. Tamara Hamilton, an animal welfare advocate in Washington D.C., argues that emotions are important in animal activism *"because we have our hearts involved."* Court points out, however, that the treatment of animals is such an emotional issue that the strong emotions can drive people in the movement apart. Put differently, there is a price to be paid for an activism that is motivated by strong feelings of compassion for beings outside most people's moral community. For example, the activist leader of the prominent English animal rights organisation Animal Aid, Andrew Tyler, was asked about the costs of animal activism. He responded in personal as well as more general terms:

> *Oh definitely, it consumes me. . . . one of the issues . . . which is neglected is the tremendous toll this thing plays upon our physical and emotional health. People suffer, people get to build up anger and massive depression, they turn it in on themselves, they turn it on their families, they turn it on the groups, which is one of the reasons we have so many splits and feuds within the groups is displaced anger and rage.* (Interview 1996)

Tyler's words highlight the price that people pay for being creatures of conscience. Time and again, grassroots activists worried about the guilt that they felt in not devoting more time to the animals. Court, one of the oldest animal protectionists in the sample and an active campaigner for the past two decades, commented:

> *I mean I always have this worry that I don't give my whole time and heart to it. . . . My professional life has always been extremely important to me, and I usually put work first. I work with children in the courts and that has always taken priority; and at the back of my mind is this awful feeling that I ought to be giving my entire life to animal rights . . . and I've usually found that I've got in the train with my briefcase, and I've gone to work, and I worry because I forget the animals for half an hour to an hour.* (Interview 1996)

Activists engaged in the resolution of social problems frequently complain of burnout or a state of emotional fatigue. Tyler and Court's involvement in the animal movement had become an obsession after more that twenty years of grassroots activism. The contrast with Holly Hazard, Tyler's counterpart in the Washington D.C.-based Doris Day Animal League (DDAL), could not be starker. Hazard says that she loves animal advocacy work but is keen to maintain a balanced life:

> *I am deeply committed and believe in this fight that I'm fighting. And the fact that I'm able to make a wonderful salary and support my family with it is just icing on the cake. But that said, you know, I go home to a husband who's a bureaucrat and two small children and a house in the burbs, and go out to the mountains on the weekends and I don't—I have not been subsumed by the frustration of having to deal with this day after day.* (Interview 1996)

Tyler, Court, and Hazard, as leaders of their respective animal protection SMOs, differ in the way that they see their roles. While all three have demonstrated a strong commitment to the animal cause, Tyler and Court are animal rights activists, while Hazard is perceived as an animal welfare advocate. One philosophy invites misunderstanding and apprehension: "*We are mocked, we are called extremists and mad people"* says Tyler. The other, represented by household names like the RSPCA in the United Kingdom and Doris Day in the United States, suggests respectability and normalcy. Court has referred to such animal welfare groups as *"people who want to play it safe."*

Having compassion all the time can be self-defeating for some. For others it is a quality that means little for animal welfare in the absence of entitlement. Yet compassion and its correlates, empathy, pity, and sympathy are the foundation stones of the animal movement. As we will see in this chapter, these emotions were strong motivations for the founders of the World League for the Protection of Animals (WLPA), Project Jonah, and to a lesser extent, the Australian Vegetarian Society. People learn compassion for animals in different ways, such as the experience one of the interviewees, Lisa, had with her cat:

> *I was in a small flat. He was on his own during the day, and then when I got home in the evening I noticed how lonely he was. Then a few weeks after getting him, he came down with the cat flu, and it sort of struck me that these animals suffer just the same as we do and that was the turning point.* (Interview 1992)

Most of the interviewees told similar stories of how they have learned from animals. For many animal protectionists, "knowing" an individual animal was the catalyst for their involvement in the movement. Individuals like Lisa and Tamara Hamilton became incensed when they discovered what was done to animals in the name of science. Like some of the leaders of the organisations profiled in this chapter and in the rest of the book, they either joined existing animal protection SMOs or started their own to do something for the cause.

ANZFAS: A UNIQUE FEDERATION OF ANIMAL SOCIETIES

Soon after the establishment of the Singer-inspired Animal Liberation branches in the late 1970s, animal protectionists realised that they needed a coalition of animal groups if they were to be effective in a country that was geographically large but demographically small. In 1980, the Australian Federation of Animal Societies was proposed as the new organisation to coordinate the activities of the animal movement nationwide. About fifty animal societies were invited to join the federation, which held its inaugural meeting at Monash University. Singer, who was and still is employed by the university, chaired the meeting. About 22 delegates and a number of observers attended, including people from the state RSPCAs. RSPCA Australia had declined to join the new organisation, which in 1984 became the Australian and New Zealand Federation of Animal Societies (ANZFAS). At that time, the RSPCA became an associate member. Although the RSPCA did not send an observer to the inaugural meeting, Richard Ryder, from the parent body in the United Kingdom, attended as a guest speaker. After the delegates accepted the draft constitution, the new organisation divided its activities into four divisions that have remained virtually unchanged in the federation's current incarnation, ANZFAS: livestock, domestic pets, animals in research, and wildlife. Two other divisions—animals in sport and entertainment, and nonindigenous feral animals—have since been included in ANZFAS's divisional structure. Late in 1996, ANZFAS changed its name to Animals Australia to distinguish the Australian operations from New Zealand's. I refer to ANZFAS when it is appropriate to include both countries and to Animals Australia when only the Australian operations are relevant.

Pragmatism in the Suites of Animals Australia

All of the Animal Liberation branches in Australia belong to Animals Australia, and their members constitute the largest proportion of individual members in the SMO. Headquartered in Melbourne, with member societies in every Australian state and territory, the organisation is coordinated by a full-time paid director assisted by two or three part-time salaried employees and an executive committee of about 16 members. Since 1986, membership has been available to individuals in addition to societies. There are now about 500 private fee-paying members in Australia, of whom about 80% are women. Singer is ANZFAS's president and a regular contributor to the SMO's quarterly magazine, *Animals Today*. In one recent edition, Singer observed how ANZFAS's success as a regional federation often intrigued people in the United States: "There has been so much suspicion and hostility between animal protection groups in the US that many of them won't work together" (*Animals Today* February–April 1995:11). ANZFAS's organisational structure is therefore of considerable interest to animal advocates and activists inside the worldwide animal protection movement.

Some discussion of the SMO's structure and how it uniquely combines the work of animal advocacy and nonviolent animal rights activism follows. But first, Animal Australia's size and position in relation to other animal protection groups is summarised in Tables 5.1. and 5.2.

The five categories listed in Table 5.2 cover a broad spectrum of animal protectionists from welfarists (RSPCA) and conservationists (Project Jonah) to the more radical animal liberationists and antivivisectionists. As the umbrella organisation for about 35 animal societies, Animals Australia is the only regional federation of its kind in the world (Oogjes & Page 1996:5). The main member societies of Animals Australia are outlined in Table 5.2.

Although the organisation is small, it has achieved a high profile in its advocacy from the suites. For example, it has made regular submissions to the Senate Select Committee on Animal Welfare between 1982 and the present. One of its strengths lies in its ability to produce credible research on the wide range of issues with which it is concerned. One prominent and respected movement leader, Henry Spira, has praised the organisation for providing "rational, factual, and meticulously documented" recommendations to the committee, adding that

Table 5.1
Animal Protection Groups in Australia

Animal Australia's member societies	33,000
Animal Australia's individual members	500
Royal Society for the Protection of Animals (RSPCA) Australia	30,000
Main Conservation and Animal Protection Groups[(a)]	
World Wide Fund for Nature (Aus.)	38,000
Australians for Animals	8,000
Australian Koala Foundation	10,000
NSW Animal Welfare League	5,000
Wildlife Information and Rescue Services (WIRES)	1,100
Australian Wildlife Protection Council	500
Other	900
Total: approx.	**127,000**

(a)Listed in *Green Pages 1996* (ACF) and the *Directory of Australian Associations March 1996–July 1996* (Information Australia, edition 25).

ANZFAS's standards "could be profitably emulated by animal lobbyists around the world" (*The Animals' Agenda* July–August 1989b:20–1).

For an organisation whose member societies embrace the entire gamut of animal protection philosophies, ANZFAS often has to walk a tightrope on controversial issues within the movement. Critics of moderation find ANZFAS's pragmatism untenable, and even the Senate Select Committee had doubts about the organisation's authenticity on the issue of rodeos. ANZFAS told the committee that rodeos were cruel and ought to be banned immediately, but if this were not possible it would accept several reforms as a compromise. The committee then suggested that this was tacit acceptance of rodeos. Oogjes, the director of ANZFAS, responded by invoking pragmatism in defense of the organisation's apparent ambivalence: "We are just living in a practical world, and we do not expect that rodeo will be stopped tomorrow, although it is our belief that it should be" (Oogjes 1991:14).

For many people within the movement, this kind of pragmatic accommodation to the status quo is unacceptable, and it is the source of much internal friction between and within groups. The leadership of ANZFAS is very much aware that its style of advocacy in the suites and on parliamentary committees does not satisfy everyone, but claims that its success is due to its practical recommendations and the membership's ability to accept different strategies to reach an agreed end point. The director claims that the "maturity in our approach . . . the need to work cooperatively with research institutions and other animal users who hold philosophies so much different to our own . . . is probably the key to the longevity and success of ANZFAS" (Oogjes & Page 1996:5).

ANZFAS, like other animal protection organisations, must frequently decide

Table 5.2
The Main Member Societies in Animals Australia and Number of Members in 1994

Animal Liberation Groups			
Animal Liberation	NSW	1,400	
	VIC	900	
	QLD	800	
	WA	370	
	SA	300	
	ACT	130	
			(3,900)
Humane Societies and Animal Rights Groups			
Animal Rights and Anti-Vivisection		1,000	
Humane Societies		750	
Choose Cruelty Free Ltd		1,300	
			(3,050)
Animal Protection Societies			
Fund for Animals		20,000	
IFAW (Australia)		5,000	
World League for the Protection of Animals (Australia)		400	
Other		150	
			(25,550)
Conservation Groups			
Project Jonah (Victoria)		n/a	
Kangaroo Protection Cooperative		200	(200)
Vegetarian Societies			
Vegetarian Society of Victoria		300	
			(300)
		TOTAL	**33,000**

between an animal welfare reformist approach and an animal rights abolitionist stance in its advocacy work. Either way, the choice it makes will not satisfy all of its members and supporters. ANZFAS knows that for animals to be "pitied but not entitled"[1] is unacceptable to its supporters. Ideologically, ANZFAS follows the pragmatic *liberationist* philosophy of its president, while programmatically, it sees itself as a *welfarist* organisation responsible for reducing as much suffering to animals as possible. While moderation is its preferred strategy and political lobbying is its main tactic, this has not always been successful, as Singer explains:

> When I first became involved in the Australian movement, in 1980, one of my prime goals was to do something for battery hens. . . . ANZFAS has been working for several

years to get a better deal for hens. Our goal, of course, is the abolition of the battery cage; and following our submission to that effect to the Senate Select Committee on Animal Welfare, we were encouraged by their recommendation that cages should go once alternatives had been shown to be feasible. Alternatives are feasible, right now, and so we began discussions with relevant government bodies, and through them with egg producers. (*Animals Today* November 1994-January 1995:5)

Singer then explains that ANZFAS's efforts appear to have been in vain, and therefore political lobbying on this occasion has not worked. Elsewhere, he points out how he and the director, Oogjes, resigned from the Hen Housing Review in protest. The RSPCA, however, chose to remain. This prompted Singer to criticise his more conservative colleagues. He also extolled ANZFAS as the stronger, more independent body and as "the real voice of the community of animal welfare organisations" (ANZFAS Annual Report 1994–95). Singer does not say how ANZFAS proposes to enact that voice, but he does see the best hope for the hens in New Zealand, where the RNZSPCA has called for a referendum on the banning of eggs produced by battery hens (*Animals Today* November 1994-January 1995).

Although the organisation has only one or two full-time staff members, the range of its activities is impressive. As mentioned earlier, ANZFAS grew out of the Australian Federation of Animal Societies, which Townend, Singer, and others had set up in the early 1980s. The federation was needed if there was to be any hope of a unified approach to animal protection in a country so large. Animals Australia has documented animal exploitation in Australia in all of its divisions: farm animals, companion animals, animals in research, animals in sport and entertainment, wild animals, and nonndigenous feral animals. Its magazine *Animals Today* reports quarterly to members on the organisation's advocacy work in these areas. Details of its style of advocacy are given in Chapter 8. For the moment, I look at the advocacy work of a selection of the organisation's member societies.

ANIMAL AUSTRALIA'S MEMBER SOCIETIES AT WORK

The unique feature of Animals Australia is its structure, which allows animal protectionists to employ conventional lobbying from the suites alongside direct action in the streets to achieve their goals. Yet the distinction between activism and advocacy, as explained in the previous chapter, is blurred in Australia. Because the movement is so small, Australian animal protectionists cannot afford the luxury of a division of labour between advocates and activists which is characteristic of the movement in the United States and, to a lesser extent, in the United Kingdom. The prototypical Australian animal protectionist, as already indicated, represents a hybrid advocate/activist whose activities encompass both indirect and direct tactics and strategies. In the rest of the chapter, I discuss the everyday animal protection work of these advocate/activists in three organisations drawn from Table 5.2.: The World League for the Protection of Animals

(WLPA), Project Jonah (Victoria), and the Australian Vegetarian Society (AVS). These have been selected because they are representative of the worldwide animal movement in that they have sister or host organizations either in the United States (Project Jonah), the United Kingdom (WLPA), or both (Vegetarian Societies). Furthermore, they represent three of the five categories that constitute Animal Australia's main membership groups, namely, general animal protection, conservation, and vegetarian societies, respectively. The remaining groups, animal liberation and humane/animal rights societies, are described elsewhere in the book.

A Glimpse of Hell: The World League for the Protection of Animals (WLPA)

The phrase "a glimpse of hell" is from a speech made by the cofounder of the World League for the Protection of Animals (WLPA), Kate Deighton, when she spoke at an antivivisection congress in London in 1909. WLPA was founded in 1898 and formerly organised in 1900 as one of the leading antivivisectionist SMOs in the United Kingdom. On 4 October 1928, it inaugurated World Day for Animals, an event that is still marked by animal protectionists throughout the world. An Australian branch of the league was formed in 1935 and registered as a charity in 1937. For the past sixty years, the WLPA branch in Sydney has continued the original campaign against vivisection—*"a glimpse of hell"*—as well as taking on a large number of issues, ranging from domestic cats in Australia to endangered wild animals elsewhere. In 1984, the Sydney branch was itself threatened with extinction when a meeting was called to discuss its future viability. One of the individuals present, Joan Papayanni, was disappointed to learn that the Sydney branch was threatened with closure. After hearing about the early history of the parent branch in London, she decided that the organisation was worth supporting.

Papayanni is now president of the WLPA, which today has a small but committed membership of about 500 people. Papayanni was asked at the meeting if she would help get the branch back on its feet by volunteering her time and energy to the cause she believed in. A small sum of money from a bequest helped WLPA reestablish itself in Sydney, and within a year, Papayanni was heavily involved in the group's activities as president.

Like most animal protection groups in Australia, volunteers and one or two paid part-time enthusiasts run WLPA. Keeping the organisation going financially is always a major worry for small nonprofit lobby groups like the league. A bequest of $200,000 a few years ago enabled it to employ Margaret Bowman as office coordinator on a part-time basis. Both Papayanni and Bowman put in many hours at home, answering appeals for help, talking to the media, and so on. Papayanni said she thought animal advocacy work was more demanding than regular work, which for her has been, at different times, secondary teaching, social work, and librarianship. Bowman previously worked as a secretary before graduating in Arts from Sydney University and then working for the Elizabethan Theatre Trust. She offered a revealing perspective on her work in animal protection:

It can be a very difficult area to work in because of the politics involved and lack of money and the world outside and trying to get yourself heard and making people see what you're trying to say. And sometimes you even lose friendships over it, and your family is against you too. You know this can happen, but I think the joy or the catharsis in the end does outweigh it. You think every year, at the end of every year when you produce your Annual Report, you feel "yes, it was worth it." And you say, "I've got another year of it." And you just keep going until I think eventually you either burn out or something will happen to make you leave, but it hasn't happened yet, not for me. (Interview 1997)

The Annual Report 1995–96 contains a large number of animal issues, ranging from the local problems of dog and cat curfews to the global issues of steel-jaw, leg-hold traps and dancing bears. The report lists about 150 life members and a small committee that runs the organisation. All of the four committee members pictured in the report are married women whose professional careers were in nursing, teaching, and acting. The profile of Bowman, who is pictured along with her cat, reveals that she came to Australia in the 1950s with her Russian-born parents. She has lived here ever since, working in various capacities, including stints as a musical performer and as an actress. She has a long-standing and passionate interest in acting.

Bowman uses her acting skills nowadays to promote animal welfare causes in street theatre and in an anticircus play with Animal Liberation and the league called, "No Place for Animals." For her, job satisfaction includes self-development and the building of confidence in her abilities. Speaking up for animals, she says, has even taught her to be even more courageous, an experience of several Australian interviewees in the study. On one point, Bowman's views are different than other animal protectionists I interviewed. For example, in the United Kingdom, Animal Aid's Andrew Tyler believes that many people in the movement are driven to despair by the experience of witnessing indifference to cruelty to animals. Bowman's way of dealing with these problems is to see, metaphorically, the "heavenly" rather than the "glimpses of hell" in animal protection work:

I think having compassion all the time is very exhausting and a very wearing thing. I think you have to have a kind of joy in discovering things, and you have to have a joy and love for animals which does include compassion. But there's a lot of enjoyment as well. It's a kind of a joy in certain rewards and results that you get sometimes for the animals. And plus you have your own growth from it as well. This is the way to experience life at the raw edge of things. (Interview 1996)

It is revealing to compare Bowman's experience of animal advocacy with Hazard's at the Doris Day Animal League (DDAL). Hazard explained in the interview that she gets "a wonderful salary" for a job that she loves. Unlike Bowman, she is paid as a full-time advocate for dealing with the cruelty of animal abusers; but like Bowman, she has learned to distance herself from the day-to-day horrors of leg-hold traps and the like:

I have a very normal middle class life. . . . I love doing what I do. People talk about being a deep sea diver or being an astronaut or being an entertainer and saying, "I get to do this and I get paid for it." That's the way I feel about what I do. . . . I have not been subsumed by the frustration of having to deal with this day after day. (Interview 1996)

"Having to deal with *this*" is Hazard's phrase for what the founder of the World League for the Protection of Animals, Deighton, called "a glimpse of hell." At the turn of the century, Deighton was referring to vivisection and to the fate of cats and dogs at the hands of the experimentalists. As we will see in Chapter 9, as the longest-running campaign in the animal protection movement, antivivsection continues to be "the most difficult issue" for mainstream animal protection SMOs. The opposition to vivisection, now as in the nineteenth century, relies heavily on the support of cat and dog lovers. It is with this in mind that the staff of mainstream animal SMOs like to be pictured with a companion animal, such as Bowman with her cat in WLPA's Annual Report, and in a recent edition of the DDAL's magazine, the founder, Doris Day, with her puppy. Companion animals were, and still are, indispensable for promoting the cause of animal liberation. For as Keith Thomas argued, pet keeping among the urban middle classes in the early modern period encouraged them to form optimistic conclusions about animal intelligence, sagacity, character, and *individual* personality (1983:119). I have italicised *individual* so as to emphasise the importance to animal protectionists of the welfare of animals as individuals. As we have seen already, the concept is basic to the abolitionist position of Regan's animal rights followers in the movement. It is their insistence on the rights of individual animals that puts them at odds with environmentalists who are interested in protecting species of animals rather than protecting individuals. Animal liberationists of Singer's persuasion are, however, more pragmatic in balancing the rights of species versus the rights of individual animals. Just as the leaders of the nineteenth-century conservationists and animal protectors joined forces to save the koala, today's animal liberationists in WLPA work closely with the Kangaroo Protection Cooperative to save the kangaroo from commercial exploitation. Again, this can be explained as a function of the size of both movements in Australia, where animal protectionists do not have the luxury of identifying themselves as *either* conservationists *or* animal protectionists.

The World League for the Protection of Animals began in England, the birthplace of the animal protection movement worldwide and the country that has done the most to spawn organisations devoted to animal welfare. The early movement organisations that began in England, and were later to serve as models for animal protectionists elsewhere, focussed mainly on antivivisection and the welfare of domestic animals. Wild animals did not enjoy the same protection as their domestic kin did until much later. For example, the Royal Society for the Protection of Birds (RSPB) was founded in 1889, nearly half a century after the RSPCA was. In Australia, however, some introduced and indigenous fauna did have legislative protection in an act of 1893 in New South Wales that conferred absolute protection for certain native and exotic species; this was the last piece of legislation to specifically protect species introduced to Australia (Walker 1991:18). Walker points out that some Australian native animals fared better than others. For example, the emu was protected well before the kangaroo, which still had a price on its head in 1866, the year that the kookaburra won protected status. According to Walker, one member of Parliament observed that the "laughing jackass," as the kookaburra was called, was protected without law, presumably because no Australian would consider killing the endearing songbird. Other wildlife, such as koalas and whales, were not so fortunate.

In the following case study, the "save the whale" campaign is described as one of the first conservation issues to achieve political importance in Australia.

"Help Me!" (Project Jonah's Advertisement on Behalf of Whales)

The history of whaling in Australia is as old as the history of European settlement itself. Soon after the First Fleet arrived in 1788, the English whalers began to plunder the rich bounty of the Southern Seas. American whalers also worked the seas off South Australia and in the first twenty years of the nineteenth century managed to kill 150,000 southern right whales (Lines 1992). Although this species of whale has been protected since the 1930s, their numbers have never recovered from the initial onslaught, when about three hundred whalers were plying their trade off Australia's coast.

At its peak, the whaling industry had as many as seven hundred ships active in Australian waters. By the beginning of the twentieth-century, the industry had accounted for the death of close to two million whales. By the mid-twentieth century, the enormous blue whale was close to extinction, and the hunters turned their attention to much smaller whales, such as the minke whale. By the 1960s, however, the industry had become its own worst enemy by overkilling to the extent that by the next decade, only one whaling station remained operational in Australia. And this station at Albany in West Australia was closed due to pressure from the antiwhaling lobby that Project Jonah led. The bloody history of whaling in Australia is told by Jenny Talbot in her handwritten, illustrated book *Whales* (1981). She explains how the international whaling industry headed by Japan continued to "harvest" whales using the latest technology, such as grenades, explosive harpoons, and survey planes.

It is the cruelty of this "high-tech" slaughter that unites so many environmentalists and animal rights activists against the whaling industry. Referring to the use of sonar, which Australian whalers have employed since 1970, Project Jonah vividly describes the whale's suffering during this "harvesting" period:

> Even if a whale dives the whalers can trace its movements as it twists and turns, frantically trying to escape. When, short of breath, the whale surfaces it is at the mercy of the Master Gunner, dispenser of a 60 kg harpoon with explosive head. The harpoon pierces the skin, exploding in the stomach. Death sometimes takes several minutes and a second "killer" shot is needed to end the whale's agony. During the death throes blood mist is exhaled from the blowhole signalling the mounting haemorrhage of the lungs, giving way to a broad low cascade as breathing becomes more laboured. If the whale has recently been feeding it will vomit squid. (Project Jonah 1977:44–9)

Spectacles of suffering, such as the one described here, enrage animal rights and environmental activists who believe that the International Whaling Commission (IWC) is nothing more than a "whalers' club" and consequently is quite indifferent to the plight of the animals. Talbot, for example, is highly critical of the IWC, which she accuses of presiding over the destruction of one species after another since it was set up in 1946. She argues in *Whales* that the IWC had no real power to stop the slaughter of the animals since the leading whaling nations could veto any measures that threatened their interests. Furthermore, Talbot believes that the overly mechanistic, regulatory attempts to save whales that the IWC uses are no substitute for the respect that "their intelligence and their loveliness" deserve. She became convinced that people's moral indignation would be aroused to protect the animals once they understood what marvellous creatures whales were. An organisation would be needed to get that message across to the public, especially to children, who being fresh to the world are not "hardhearted" and have a "feel for injustice" (Talbot (1981:174).

Project Jonah was set up in 1971 in the United States by Joan McIntyre to teach children about valuing wildlife. An Australian master's degree scholar, Joy Lee, was studying in America and was drawn to the campaign. She later set up the first branch of Project Jonah in Sydney. In mid-October 1976, Project Jonah held its first meeting in Victoria at the home of Talbot, who describes the group's genesis in her book *Whales* (1981), which she dedicated to Lee. Late in 1976, at the same time that Animal Liberation began in New South Wales, Project Jonah held its first public meeting in Victoria, some five years after the group had been established in America. Like several other animal rights activists I interviewed, Talbot believes that she was "called" to save the whale. In her case the fateful moment came in a "great dream" in 1970:

> *I knew it was something of absolutely major importance. I've actually never had a dream where I've been touched before, except that dream. I felt the whale's velvety touch on my heel deep in the water, this green water, and I was walking on the water. Isn't that shocking? That's a sort of Jesus Christ, some sort of insane "save the world" complex*

or something. I don't know what it is, but I was walking on the water with my best friend and we were pushing our prams along. . . . and I actually had, really I would have to say, the truest sense of being called in my whole life, during that time, during that, whatever it was sixteen, eighteen months. (Interview 1996)

In *Whales*, she describes the same experience and how it propelled her on a quest to study whales, then to paint them, and eventually to campaign against the killing of the animals for profit. Talbot's story of the founding of Project Jonah in Victoria is important for what it reveals about the meaning and politics of grassroots activism for an individual with a cause. Like several other people in this study, Talbot believes that she was "called" to defend threatened animals. When she decided to do more than prepare art exhibitions on whales, she searched in vain for like-minded activists in the local animal welfare and environmental fraternity. Friends of the Earth, for example, had no "save the whale" campaign. She eventually heard about a woman in Sydney "doing something about whales" and contacted her for the first time in 1976. Later that year, the woman, Lee, came to see Talbot and provided her with a list of about two dozen people who had expressed interest in Lee's radio program on Project Jonah. Lee urged Talbot to set up a branch in Melbourne and to prepare for a protest against the IWC, which for the first time was to meet in Canberra in June 1977. Talbot believed that she had found her calling and was the right person for the task:

And Joy, of course, when she came and she saw my paintings and heard what I'd been doing, she knew she'd found her person. She wanted the right person, and she found me, and it's true and I was. (Interview 1996)

After the inaugural meeting at her home in October 1976, the group had its first public meeting at the end of the year. Talbot describes how the *"tiny, little, freaky movement"* became a huge success after they placed their large ***"Help Me!"*** advertisement in *The Age* newspaper. Donations soon came pouring in, along with a vast number of letters of support, to help Project Jonah achieve its goal to "save the whale." Talbot's documentation of the next several months, in *Whales*, reveals how the group's activities peaked during the visit of the IWC in Canberra, when whaling in Australia became the most important environmental issue on the political agenda. According to Talbot, this was the first time that an environmental/animal welfare issue had captured the public's imagination in Australia, so much so that it became a federal election issue in 1977.

Talbot points out that Project Jonah, at the time, was not aligned with any political party or religion and that people of all persuasions joined the group. Ironically, it was the grab for power inside Project Jonah that effectively ended Talbot's involvement in the fledgling SMO. The sudden popularity of whales in the media attracted the interest of individuals who could see the political potential of Project Jonah:

I learned that the movement from a tiny, little, freaky movement to a huge success . . . it didn't matter that it was environmental and not BHP [Broken Hill Proprietary Limited]. As soon as you've got something going, there's six people who want to grab it and use it for their own ends. (Interview 1996)

Talbot (1981) wryly notes in *Whales*, "Every mouse has its day and I'd had mine." But although her committee work had ended, her passion for defending nature continues to this day. In the conclusion to *Whales* she ponders:

> Will we humans only be content when there is no wilderness left, no animals that are truly "other," when all plants and animals that exist exist because they are "of use"? Will we live in plastic-domed cities, surrounded by poisoned seas, drug-taking parasites hooked up to the television? (Talbot 1981: 189)

It is not surprising then that Talbot, a self-confessed Luddite, is concerned about the "evil" associated with technology such as what she calls the technological horror of genetic engineering. Moreover, she expresses an exasperation with the human race that borders on misanthropy, and she suggests that people in the animal and environmental movements think that this is neither unusual nor unwarranted.

I'm very interested in Buddhism and the Dalai Lama, and I have to work very hard at myself to think compassion, to try and think love, because I regard humans as the most vicious and evil force that's ever walked the face of the planet, and I suspect a lot of animal people and conservation people would feel the same way. Just look around you and it's destruction, destruction. (Interview 1996)

Very few people interviewed for this book expressed sentiments as strongly as these, although many would sympathise with her point of view. Talbot is atypical for another reason as well in that she is one of the few individuals in the study who sees connections, rather than disjunctures, between environmentalism and animal rights. She believes that they are linked *"as a microcosm of the macrocosm"* and that both are in imminent danger simply because humans have lost the Buddhist sense of balance and moderation; spiritual values have been rejected for the new religion of progress through technology and materialism:

Basically, I believe that there has been such an imbalance between human feeling, human knowledge, there is such an imbalance between that and technology, that I virtually think that there should a moratorium on all research because the horrors of what is happening are directly related to this huge imbalance . . . and [quoting Jung] "that the lack of spiritual values and the frustration and the not knowing have led to a universal will to destruction." (Interview 1996)

In 1972, the United Nations Conference on the Human Environment adopted the whale as a symbol of nature under siege. Project Jonah took up the challenge at the grassroots level by arguing in one of its earliest policy statements that "To destroy the whales for the short-term needs of a minority percentage of the world's population is to acknowledge that we are powerless to save the planet" (Project Jonah 1977:44). For Talbot, it is whales, rather than humans, who inspire in her the deepest feelings of compassion and respect:

> *I saw whales as almost above all other animals in being a symbol of the abominable things that humans do to plants and animals. Their intelligence and their loveliness and to see those documentaries—which I can really barely watch—with them making those little noises and screaming and then their blood, like one of the documentaries where they killed both a baby and a mother. You see the milk and the blood all over the sea and all that kind of stuff. I find it just absolutely heartrending. There's something about it that touches on my vibratory level and I think touches on a lot of people, obviously touches on a lot of people. Those little noises, there's something so wonderful.* (Interview 1996)

Sentiments of this kind, which have driven "save the whale" campaigns, succeeded in mobilising thousands of Australians to the cause. Since 1978, when Australia ceased whaling, whale watching has become a major attraction and a boost to the tourist industry. Many of the pioneer activists against whaling have now focused their campaigns on the remaining whaling nations, such as Japan and Norway. The president of Project Jonah, Geoff Mosley, claimed in a letter to the editor of *The Age* 15 May 1993 that a cessation of commercial whaling would be a symbol of a new wildlife ethic that rejected cruelty and the endangerment of species for ecologically sustainable development. It is not clear from the letter how sustainable development is possible if a ban on commercial whaling is to be implemented. Not surprisingly, "save the whale" advocates have been challenged on this and other grounds.[2]

In condemning antiwhaling activists as people driven by sentiment rather than by sanity, critics misread the new environmental ethic, which is actually more anthropocentric than pro-animal rights. This ethic, or new ecological paradigm (Catton & Dunlap 1980), is based on the idea that the current threat to species survival has been brought about by the insanity of commercial greed as manifested in the unrelenting exploitation of nature, which has wrongly been perceived as an infinite resource. The case of BSE or "mad cow" disease in the United Kingdom is one of the more recent and alarming instances of nature's backlash against human interference. While the BSE tragedy shocked people worldwide, it came as no surprise to many advocates of vegetarianism, whose ideas are the subject of the next section. I should point out, however, that while animal rights/vegetarian activists interviewed for this book were all aware of the meat scandal in England, none attempted to capitalise on the scare. And in my interview with the president of the Australian Vegetarian Society, which is discussed next, the subject never came up.

A New Age Sideshow: The Australian Vegetarian Society (AVS)

Although there are only approximately two thousand members of the AVS, the numbers are steadily growing, according to its president, Mark Berriman, who runs the society from his private residence in inner-city Sydney. Berriman was formerly a spokesperson and onetime president of Animal Liberation (ACT) in Canberra. Now in his early forties, he has been a vegetarian since 1981. At the time of interview, he was heavily involved in setting up an AVS stall at the annual *Mind, Body, Spirit* festival in Sydney. According to Berriman, this "New Age sideshow" has attracted a lot of interest over the years. People were at first sceptical or indifferent about vegetarian/animal rights propaganda but now take the advocates more seriously, not simply because they persist in setting up their stalls each year but also because public sympathy for vegetarianism has grown. To make the point, he cites the words of philosopher John Stuart Mill about all great movements going through three stages: ridicule, discussion, and adoption.

Berriman is optimistic about the future of progressive movements like animal rights, environmentalism, and vegetarianism because he believes these are genuinely emancipatory movements that are attractive, especially to young people. He explained that his own conversion to vegetarianism was for aesthetic reasons and believes that this is what motivates many of today's new converts as well:

> *And I think similarly with the younger people, there's a certain aesthetic change that meat eating is no longer necessary. It's barbaric, it's a bit dinosaur, like bear-baiting or something like that.* (Interview 1996)

Berriman expresses sentiments similar to Talbot's about the values underlining modern technology. He sees an enormous gap between advances in communications technology and what he calls, "our central ethics." It is a gap that he admits finding difficult to live with, but he is confident about using communications to bridge the ethical gulf:

> *The Industrial Revolution's come to its end, we're at diminishing returns as far as technology is concerned... We want a system of values that reflects the environment equally with the financial world because the two are interlinked, not the sort of act now and pay later mentality. And it all ties in, it's just a logical extension. If you get people eating lower on the food chain, well you're automatically going to be saving on your health bills.* (Interview 1996)

One of the main technologies that advocates like Berriman use to promote their cause is electronic and print media, ranging from E-mail to the conventional exchange of newsletters and movement magazines. The AVS, like most SMOs described in this book, is linked up to a network of organisations via the Internet and e-mail. Berriman sees communications technology as a powerful tool for the wide dispersion of progressive ideas and for the ultimate liberation of individuals from old habits like eating meat.

One of the strategies he favours is to *"sell the same soap powder in a different package,"* by which he means promoting the positive, liberating message of vegetarianism, rather than the more disturbing themes of cruelty that animal liberationists emphasise or of ecological catastrophe that some environmentalists preach. For the AVS, which has just recently gone into a joint-publishing venture with a natural-health magazine, it makes sense in an increasingly ageing society to focus on health and lifestyle issues. Berriman argues that vegetarianism is not just about individuals improving their health, but that it has wider implications for the health of the environment. Thus, the AVS strategy is to expand the rationale for vegetarianism so as to include both personal and environmental health considerations alongside traditional animal welfare concerns.

As president of AVS, Berriman is anxious to unite like-minded individuals who want a better, more sane balance between their lives and the natural world. The AVS, which was founded in the 1930s, reached its peak in the 1950s. Today, it is trying to rebuild its support base by becoming more inclusive and by emphasising the common links between people interested in social justice issues. Berriman sees improved communications and PR as the ways to overcome the traditional conflicts that have divided kindred spirits in the past:

> *It really is important to get that communication going, and just a respect for the other person, a strong respect for the other person's point of view or whatever. In situations where it seems insoluble then, just keep the debate going and just with the will on both sides, it will resolve itself.* (Interview 1996)

Like many leaders of new social movements, he realises that forging what appear to be *"logical, natural, inevitable connections"* between people in new social movements—animal rights, natural-health, peace, women's, and environmental movements—is no easy task but must be accomplished.[3] The alternative, he suggests, is that *"we'll all go down the gurgler together."* He adds:

> *The same motivation that makes you fight for an animal means you can liberate a political prisoner of some other country, or whatever. And if you're in a peace movement, well how can you have peace if you were sitting down to a fund-raising dinner using slaughtered animals?* (Interview 1996)

Adams (1990) poses the same contradiction for feminists. She chastises those who are "trafficking in animals," by which she means those feminists who attend conferences and dine on animal flesh at the meal breaks without considering the implications. For Adams, the consumption of animals represents a form of oppression that women presumably would want to oppose as feminists. Kathryn Paxton George (1994) asks the broader question, *Should Feminists Be Vegetarians?* and responds in an article of that name in the negative. George, who is a practising vegetarian, argues that the nutritional needs of women are different from those of men and vary at different stages of the life cycle. Needless to say, George's paper stimulated a debate in which a number of prominent feminists, including Adams, responded with various counterarguments.

In his advocacy of moderation and solidarity with other progressive movements, Berriman is clearly promoting context-sensitive vegetarianism of the kind advocated by George, rather than the zealous abstinence demanded by fundamentalists in the animal rights/vegetarian movement.

I asked him to comment on what he thought could be a model for social movement organisations associated with vegetarianism, animal rights, and the environment. He singled out ANZFAS's regional structure, which allows accessibility and input, as one of the best models "for organising and sharing ideas and working together." Berriman believes that animal rights/vegetarian advocacy work in Australia is more promising than it is in either the United States or the United Kingdom, where he suspects, increased corporatisation and bureaucratisation have diminished activists groups' effectiveness. What Australian groups lack in size, he suggests, is compensated for by their grassroots egalitarianism:

> *I think we're small, we're highly organised, ideas can move much more quickly than they can, say, in the older countries, certainly in Europe. We're sufficiently introspective—it's our cultural cringe if you like—to take on ideas and assess them a bit more carefully before we work out which ones we're going to accept. So I think that does give us a unique perspective.* (Interview 1996)

Ironically, in a country where distance is often seen as a national handicap, Berriman sees no such obstacles to communication via the small, activist networks throughout the country:

> *I think you can still get the ideas out, there's a lot of word of mouth that goes on. I think more than anything it's word of mouth. Someone comes across an idea and spreads it . . . So it's not necessarily size. And people are traditionally, I think, suspicious of size too. They think, is it an organisation that's being supported for the organisation's sake or what it actually represents? And I think we have a natural suspicion of that as well in our egalitarian past.* (Interview 1996)

Like most other Australian animal protectionists, Berriman is sceptical about large-scale, professionalised advocacy groups of the variety that have developed in the United States and, to a lesser extent, in the United Kingdom. In a society where cutting tall poppies down to size is a national pastime, egalitarianism of the kind that Berriman advocates is an idea that has wide appeal. However, this grassroots style of Australian animal protectionism, based on informal "word of mouth" between enthusiastic amateurs, has its own limitations. These will be taken up in the final chapter. In the next chapter, the meaning of grassroots activism—*the heart of the movement*—and its links to ethical vegetarianism are examined.

NOTES

1. I have borrowed this apt phrase from the title of Linda Gordon's study of single mothers, *Pitied but not Entitled: Single Mothers and the History of Welfare*, Cambridge, MA: Harvard University Press (1994)

2. A month before Mosley's letter was published, a well-known columnist from *The Australian* newspaper criticised "the anti-whaling lobby"—a term that is designed to have less public appeal than "the save the whale lobby"—for putting sentiment above sanity. The columnist, Paddy McGuinness, argued that the demand to ban whaling was based on the "moral and cultural imperialism" of "the lunatic fringe, the hug-the-whale types" who deceive people with "animal liberation propaganda." McGuinness, an apologist for the commercial whaling nations, further criticised "the propaganda of the professional environmentalists who live well off their fundraising activities, and who feel a kind of wooly sentimentality about whales" (*The Australian* April 1993b). At the time he wrote the two articles, McGuinness was a guest of the Japanese government, which he praised for having accepted a degree of international supervision that countries such as Australia would not be prepared to tolerate. On that point he is undoubtedly correct.

3. I have taken a similar position in suggesting how coalitions might be made between such groups to expose the dangers of toxic wastes and their links with animal experimentation (Munro 1999a).

Chapter 6

Grassroots Activists and Ethical Vegetarianism

And then someone threw a brick through a window!

—English animal rights protester, 1996

In this study of the animal protection movement, two broad types of social movement activity have been identified: grassroots activism and organisational advocacy. Grassroots activism typically occurs when animal protectionists engage in direct action on behalf of animals, often in actions not associated with formal organisations, and usually as unpaid volunteers. By way of contrast, animal protection advocates are defined in this study as individuals who run their campaigns from the suites as paid employees of animal organisations. Interviewees in the study had little difficulty choosing the designation that best matched their involvement in the animal movement, with the number dividing evenly between activists and advocates. While the distinction between movement activists and advocates is often more blurred in the real world of social movement activity, the ASIS sample (Munro 1996a) of animal protectionists in Australia also identified the broad orientations as follows: activists (16 per cent), advocates (33 per cent), supporters (46 per cent), and "animal lover" or a combination of the three main designations making up the remainder (5 per cent).

An Australian activist, Casey, distinguished between activism as "hands on" and advocacy as "behind the scenes." This corresponds to the streets/suites designation I use in the book:

I think the campaign side, you're working with a lot on paper, the media side, the political side, whereas the activist side, you've basically got one aim and that is to get out there and help the animals at whatever cost. Activism is hands on, whereas the other [advocacy] is, as I said, behind the scenes. (Interview 1994)

While grassroots activists like Casey see the work of their advocate counterparts in the suites as important, others disparage it as just another day at the office. Activism in the streets is less constrained, and many activists believe that it avoids several inevitable pitfalls of organisational advocacy work (Flacks 1988: 196–7). Although Flacks's focus was on the American Left, what he describes as "the dilemmas of organization" (1988:193) are familiar to all social movement activists and advocates. Essentially, many activists reject the bureaucratic, hierarchical, organised professionalism of the suites for the spontaneity of the streets. Even so, I argue that both professional advocacy and amateur activism are needed for building a movement strong enough to defend the welfare of animals.

BECOMING AN ACTIVIST

The activists I interviewed all had different conceptions of what it meant to be an activist rather than an advocate. That they defined themselves as one rather than as the other was partly determined by what motivated them to take up the animal cause in the first place. No attempt was made to probe the interviewees' psychological motives for activism, as attitudinal explanations are generally held to be of limited value in explaining participation in movements (McAdam & Paulsen 1993:643). A more sociological approach is suggested by Stryker's (1968) concept of "identity salience," which holds that activists are recruited to participation in stages: first, the individual experiences a recruiting appeal that links the movement with the salient identity (e.g., animal lover, champion of the underdog, vegetarian, environmentalist, radical, etcetera); and second, there must be confirmation from those who normally sustain the individual's identity that participation is justified and will not jeopardise other commitments or loyalties, such as in the family or workplace.

These stages fit most of the activists in the study, at least in the way that they were initially recruited, or more accurately, motivated to participate in animal rights activities. As the following case histories indicate, the recruiting appeals came in several guises, ranging from direct and indirect public appeals via the mass media to more specific experiences via the appeal of vegetarianism or books such as Singer's *Animal Liberation* (1975). As McAdam and Paulsen argue, recruitment involves a strong subjective identification with a particular salient identity reinforced by organisational or individual ties (1993). The case histories below generally conform to this pattern of activist recruitment, although there were several cases—especially Rhett, Sherry, James, and Tina—where personal relationships were subjected to tensions as a result of the individual's conversion to animal rights or vegetarianism.

Australian activists

Roger

A well-built man in his thirties, Roger describes himself as a "hippy-biker" type. Roger first got involved in animal protection when he answered a call in a newspaper for volunteers to help injured and burnt animals following the catastrophic "Ash Wednesday" bushfires in 1983. He says that the experience of saving animals in a dangerous environment gave him a sense of doing something constructive, which affected him in a profound way. Most people will condemn cruelty to animals, according to Roger, but few experience the transformative, emotional power that comes from the direct contact of rescuing individual animals:

But once people get hands-on, they become enthralled with that animal, they have a personal involvement with it. I've seen it with wildlife, with ducks. It's an amazing sight on that duck campaign when a rescuer picks up their first bird, and that result, whether it's dead or alive, it has an effect. (Interview 1994)

He recalls that the first bird he ever rescued nearly took out his eye:

I can remember the heartbeat. I can remember the calming effect of covering the bird's head loosely with a cloth its heartbeat slowed slightly, the stress level had dropped. I felt I had done something constructive, something positive to relieve the terror and the horror that bird was experiencing. I think most rescuers have similar experiences. In the other campaigns, it's usually frustration that drives you. (Interview 1994)

Rhett

A quietly spoken public servant, Rhett decided some years ago to take the step to animal rights activism. He claims it was the inconsistency in our treatment of animals that was the catalyst for his activism and the cause of some tension in his personal relationships:

But I can tell you that when I was a child, I was presented with an inconsistency which always stuck in my mind. And that was that my father would never eat fowl, and at Christmas time he would always have a chop or something like that. And the reason was that when he was a child, his father brought home some chickens. The kids had made pets of them, and then they were served up for Christmas dinner. He was so upset that he refused to participate in the killing of chooks . . . This was when I was quite young, and I could see that he was being inconsistent, but I'm grateful for the inconsistency because if I hadn't had that example of someone who was sensitive, I mean, who knows? It might have taken me ten more years or something. I don't know. (Interview 1994)

Al

As an environmentalist who is reluctant to call himself an activist, Al prefers to make "*a quiet statement*" by practising an eco-friendly, cruelty-free lifestyle rather than by participating in protests and the like. He was attracted to the animal movement by the logic of vegetarianism and the promise of moral purity:

In fact, really you should be vegan. That's part of the reason why I've tried to be vegan . . . I will not knowingly cause death or suffering to animals, but even by consuming free-range eggs or milk, you're still contributing to animal suffering because clapped-out old dairy cows and laying hens end up in hamburgers and chicken soup, respectively, and most of the male chickens and the male calves go straight to the abattoir. . . . I think to be morally pure, you have to go all the way to being vegan, which practically is very difficult in our culture. (Interview 1994)

Gaynor

As far back as her childhood, this professional woman in her thirties can remember doing things for animals, like rescuing baby birds that had fallen out of their tree or even liberating rats and mice destined for the high school dissection table. Gaynor cannot say why she has always had this inclination since she is the only one of her nine siblings to be interested in animal welfare. In fact, for a long time she felt alone and different, especially with regards to her revulsion against meat:

To me, the fact that, I mean, you eat a leg of lamb, to me, I would find the idea of eating a leg of lamb as amazing as taking a chomp out of your leg. To me, there's no difference, there's absolutely no difference. (Interview 1994)

Like many animal protectionists, she found Singer's *Animal Liberation* (1975) a consolation rather than a revelation:

It just really reinforced it, and it was good to feel that I wasn't alone. Because at that time I wasn't involved in any groups. It didn't make me feel special but it made me . . . sort of like, not holy, but it made me feel a bit pure . . . I felt different to other people, and it was reinforced when I read those types of books. I thought, "Yes, I'm not alone," because I did feel alone. You know, I'd go to my mum's and she'd say, "Are you still vegetarian?" You know, that sort of thing, as if it was a phase I was going through rather than a lifestyle I'd chosen. (Interview 1994)

Sherry

In Sherry's case, however, the effect of Singer's book was more direct. She remembers reading *Animal Liberation* in 1984 and thinking how everything "*suddenly became obvious.*" However, it was also very obvious that her husband at the time did not go along with the idea of extending moral consideration to animals, especially at the dinner table, and so Sherry had to wait until after her divorce five years later before joining a branch of Animal Liberation. After reading Singer's book, she realised how it would be possible to lead a cruelty-free life by giving up meat and rejecting animal-related products. But the fateful moment only came when she participated in a duck-rescue operation:

I've never been so devasted in all my life. It was just terrible! And from that moment, I was fully committed, more so than anything I've ever been involved in. (Interview 1994)

Casey

Casey believes in hands-on activism rather than in the *"behind-the-scenes paperwork"* that she characterises as advocacy. For her, vegetarianism is visceral: *"I can't imagine loving something and then eating it!"* She is opposed to all forms of animal cruelty but is particularly incensed by animal experimentation, especially in behavioural science. She says that the longer she is connected with the movement, the more "passionate" she becomes. Her ambition and motivation is to get involved in exposing cruelty by using video footage to save animals' lives: *"If you can change one person, you are saving hundreds of lives."*

English and American Activists

Joan

At 76, Joan is the oldest female animal activist in the sample. She says that she simply progressed into animal rights about twenty years ago without any specific incident acting as the catalyst. Later in the interview, however, she explains why she joined the recent protests against live animal exports from British ports:

Pity and anger! And a general sort of compassion, but anger really I think . . . But it's because animals have no defender, and I object to cruelty. Who doesn't? I think it's spiritually wrong too. (Interview 1996)

Joan, like some others in the sample, believed that her involvement in animal rights activism was literally a calling:

And I sometimes have this sort of pseudospiritual feeling that some of us are called. You get that feeling when you're talking to people. It's as if the animals are calling us. And you're always warned, by people who know, don't look a sheep in the eye. I discovered that too late at the market because a sheep looked at me very seriously, and I feel now we're betraying them all by not stopping this bloody trade. (Interview 1996)

Milly

The live animal exports issue is also the reason for Milly's militancy. As a forty-year-old London social worker, Milly, unlike Joan, is a first-time animal rights protester, although she has been a passive supporter for about a decade. In common with Joan, she felt that the animals were calling out to her for help:

> *[I]n fact it was quite strange really. It was a bit like a clarion call to like-minded individuals. I think people saw that there was an opportunity for the average person in the street to do something positive. And people came from all over England and Scotland too . . . I had, you know, written off, lobbied MPs, the usual things, written letters and signed petitions and what have you. But it was the first time that I had actively taken to the streets.* (Interview 1996)

Like hundreds of other people, Milly had been mobilised by the potent images on television that had been featured in the mainstream media for several months in early 1995:

> *I think that the pictures and the shots of the animals going past in the lorries, I think that's what touched me really. And I thought, well, this is ridiculous in this day and age that that type of trade should be happening.* (Interview 1996)

James

Of all the activists in the study, James was the most intense about animal rights. His long-standing opposition to hunting has brought him into conflict with the law, which has meant being arrested on about fifteen occasions. As a child, he remembers frightening away animals being stalked by hunters and being shouted at by angry men with guns. In recent times as a hunt saboteur, he claims to have been beaten up and generally "hassled" for his animal rights stance, which included antivivisection activities at the university where he works. He seemed momentarily at a loss for words when asked about his motives. He went on to give a singularly pragmatic reason for his abiding opposition to hunting, namely that it is *"the easiest target."*

Mike

Opposition to blood sports is also what drives Mike Huskisson, who is profiled in Chapter 11. As a veteran of animal rights activism for over 20 years, he too has been arrested many times and has served prison sentences for his animal rights activities. He is still driven to protest against unnecessary cruelty inflicted on animals by hunters and vivisectors. He devotes himself full-time to investigating cruelty to animals as founder of the Animal Cruelty Investigation Group (ACIG) in the United Kingdom. His early experience with hunting was the original motivation for his activism:

Well, I used to live out in the countryside near Cambridge, and the local foxhunt had two meets in the village. I used to go out and watch them, and I didn't like what I saw, and it grew from that. It started from 1971 when I went up to university in London. I then joined the Hunt Saboteurs Association and got involved in a lot of sort of direct action at the time. (Interview 1996)

Phyllis

In her sixties, Phyllis has been involved in animal welfare for about twenty years and cites Singer's *Animal Liberation* (1975) as the original catalyst for her activism. She also mentions her experience, as an eight-year-old girl, witnessing the killing of farm animals for food, which may have been the subconscious trigger that led her in that direction in the first place. Phyllis has devoted thousands of hours to animal welfare proselytising at one of the many Animal Aid fund-raising and information booths throughout England. She has become even more active in her retirement by campaigning against the live export trade, which she describes as *"this awful cruelty."*

Tina

Tina is an American university graduate with postgraduate qualifications in environmental studies. It was in graduate school that she did a course on animal rights law and began to see a different side to vegetarianism, which she had been practising for several years, not without some costs to personal relationships. The original conversion to vegetarianism had nothing to do with ethical considerations but rather with aestheticism; she explained how she had been impressed by the youthful looks of a vegetarian she had sat next to at a wedding. However, she later realised that *"I was just being completely hypocritical, with all the environmental damage that goes on."*

These dozen case histories broadly correspond to the focal concerns identified in ASIS (Munro 1996a) as the main factors that motivated people to support the cause. There were four main issues that respondents said continued to drive their participation in animal welfare: cruelty and exploitation of animals, farming practices, animal experimentation, and a general belief in the rights of animals to be left alone. While it is true that all of the above case histories specifically mentioned their distaste for cruelty toward individual animals, they usually cited a particular abuse that they found objectionable, such as hunting, practising vivisection, live exporting or meat eating. The case histories also reveal, for most of the interviewees, a "fateful moment" when they made the decision to become an animal rights activist. It was this moment that the above histories seek to highlight. These can be summarised in general terms as animal activism via vegetarianism (including the influence of Singer's *Animal Liberation*) or animal activism via the experience of saving individual animals, including the desire to save animals in contexts such as in the wild, on farms, in labs, or in the recent case in the United Kingdom, from live animal export. About half

Table 6.1
Dietary Regimes of ANZFAS Members and Levels of Activism

	Activists %	Advocates %	Supporters %	Others %
Vegans (n=50)	32	12	3	53
Vegetarians (n=221)	55	60	42	16
Carnivores (n=166)	13	28	55	31
per cent	100	100	100	100
Total	71	145	202	19 (437)

of the activists—Rhett, Phyllis, Gaynor, Mike, James, and Casey—were motivated during childhood or in their youth, while the remaining half—Roger, Al, Tina, Sherry, Joan, and Milly—came to the cause at some later stage in their lives. In virtually every case, however, they all shared a "salient identity" as an animal rights/vegetarian activist, antivivisectionist, environmentalist, or conservationist that has continued to motivate their participation in the movement. In several cases, there were tensions associated with being an animal activist with or without the vegetarian identity.

BEING AN ACTIVIST: THE VEGETARIAN CONNECTION

One of the factors identified in ASIS (Munro 1996a) that distinguished animal rights activists from advocates and supporters of animal welfare was the respondents' dietary habits. As expected, only a small percentage of activists were meat eaters. At the other dietary extreme, vegans were much more prolific among activists (32 per cent) than among either advocates (12 per cent) or supporters (3 per cent). Respondents with weaker attachments to the animal movement were much more likely to eat meat. Supporters were four times more likely than activists to be carnivores, while the percentage of meat-eating advocates was double the percentage for activists. These tendencies are shown in Table 6.1.

The conclusion we can draw from this data is that the more active the member (according to their self-designation as activist, advocate, or supporter), the more disciplined are their eating habits vis-à-vis meat avoidance. As is often the case, however, the reality behind the statistics is more complex. Of the dozen activists described above, only two are strict vegans, and another three individuals are struggling to make the transition from vegetarianism to that more pure state of eating virtuously, veganism. The meaning and importance of food to animal rights activists cannot be underestimated, for to many activists, the two issues are different sides of the same coin. For the most passionate animal rights activists, eating meat devalues the movement's currency as the champion of the rights of animals to be left alone. The true believer, therefore, sees the avoidance of meat as the most basic prerequisite to movement membership and credibility. Vegetarianism is not, however, without its costs, as Sherry points out:

I do miss it. I'll be honest. I used to love the taste of meat, and I do miss it. Every now and then, my mouth will water when I think of it. But then I look at the animals, and I think, well, basically every type of meat you've got, I used to own as a pet at one stage or another, and I just can't do it. (Interview 1994)

Realists Versus Purists: Finding the Middle Ground

Subtle propaganda is the best way to produce converts, according to Al. He says that he gets exasperated when teenage girls inquire about vegetarian makeup and shampoo not tested on animals because they cannot bear to cause suffering to animals:

You feel like saying, "You stupid girl, by switching to vegetarian lipstick, you might save one rabbit a year, but by becoming vegetarian, you can save ten bullocks and a hundred chooks a year" . . . But you'd hope that if they work here at the Animal Liberation branch for awhile and are exposed to our literature and read our magazine and get harangued by ratbags like me, they'll eventually get converted. (Interview 1994)

Yet many people in the animal movement are just as adamant as the purists in arguing for a nonvegetarian approach to animal rights. Huskisson, an activist for more than 20 years, puts the realist argument against ethical purity in these terms:

The fight really is for the middle ground of public opinion. And if they can say, "Oh these are a load of bloody nut cases that won't even wear leather shoes and drink milk" and all this sort of stuff, then it makes the ordinary person say, "Well, that's so alien to me I couldn't be a party to that." (Interview 1996)

Sentiments of this kind were particularly evident in the comments made by organisational advocates in SMOs like the League Against Cruel Sports (LACS) and the American Humane Association (AHA). Some movement leaders believed that the strict avoidance of meat and animal products was a tactical mistake when arguing the case for animal rights. For example, John Bryant, a prominent antihunting advocate for The League Against Cruel Sports, believes that puritanical thinking, particularly in the early years of the movement, was counterproductive to movement mobilisation:

This is what happens in the animal rights movement, not so much now, but it did happen when animal rights groups were being set up everywhere. It was a big issue in the 1970s and 1980s, and people were going along to meetings to find out because they were troubled. Consciences were troubling people, but they'd go in there and someone would say, "You've got leather shoes on! That's disgraceful!" and they'd be turned away and made to feel really uncomfortable, and they're lost. (Interview 1996)

Scott Williams, of the American organisation Farm Animal Reform Movement (FARM), describes himself as a vegetarian/animal rights activist. He shares Berriman's view described in the previous chapter, that veganism and vegetarianism should be used to promote a positive message, which is ultimately more attractive to people than the prescriptive animal rights manifesto of the true believers.

There is therefore a good deal of ambivalence and controversy amongst activists about the meaning of meat and about what standing carnivores have in the movement. Two questions in ASIS (Munro 1996a) dealt specifically with respondents' attitude to meat and meat eating. In terms of moral reprehensibility, respondents ranked eating meat only ninth in a list of fifteen practises involving human uses of animals. Furthermore, in a list of eight statements that included the suggestion "of all the forms of animal abuse, meat eating is the worst," respondents displayed more ambivalence on this statement than on any other issue. Not surprisingly, meat—or as the purists prefer, animal flesh—is one of the most challenging moral dilemmas for animal protectionists inside the movement. For many, the avoidance of animal flesh and animal products remains the most important act of compassion for anyone purporting to be an activist in the animal movement. To the true believers in the movement, ethical vegetarianism and animal rights activism are synonymous.

One can, however, detect gradations in virtue when it comes to the avoidance of meat. Vegetarians are seen as less pure than vegans, while carnivorous members are, at least in the United Kingdom, increasingly viewed with suspicion, as Joan Court pointed out:

Lyle Munro: Would you regard it as the probably the worst form of animal abuse the fact that food, that animal flesh, is used as food?

Joan: Yes, of course, and all that goes with it. The sort of cruelty that goes with it, the factory farming, and I think, of course, it's quite dreadful. People use the word "obscene" all the time so it seems to be dreadful to eat animals.

L M: But then, are there people in the movement who are meat eaters?

Joan: There used to be. We used to say it doesn't matter if you're a vegetarian or not. That's a joke now. Everybody I know in the movement is vegetarian drifting towards veganism. The young people can't understand why everybody's not a vegan, but we're now attracting a lot of older women and men who have had a lifetime, of course, of eating dairy products and it's more difficult for them. (Interview 1996)

At one point in the interview, which took place at the kitchen table that we shared with a cat and his open tin of pet food, Joan claimed that no meat was consumed in her household, except by the cat. As if to explain the contradiction, she observed that *"It's not possible to walk on this earth without being cruel to animals. But we try to reduce it."* The next section explains how grassroots activists seek to alleviate animal suffering.

VARIETIES OF ANIMAL ACTIVISM

While there was considerable ambivalence amongst ASIS respondents concerning the issue of meat, there was generally an overwhelming consensus on what constituted appropriate action that animal protectionists could take on behalf of animals. Illegal activities, such as liberating animals from their alleged torturers or inflicting damage to property associated with allegations of animal cruelty, were generally eschewed by the respondents in preference to legal, conventional efforts, such as media promotions and education. This seems to go against the conventional image of the animal rights activist—for example, Shapiro's (1994:148) notion of "the caring sleuth" described in Chapter 2—as a passionate exponent of aggressive direct action and militancy. Interviews of activists in this study generally confirmed his description. That a more conservative style of activism was revealed in ASIS was undoubtedly due to the nature of the questions, which did not allow the discursive opportunities available to individuals in the interviews.

Tactics in Animal Protectionism

How far are animal protectionists likely to go to achieve their goals? Although this question was not put to the interviewees, the people in the ASIS sample were asked to rank a number of tactics in terms of their justification and effectiveness for achieving improvements in animal welfare. The results were compared to data from an American survey.[1] Respondents in both countries agreed that all five of the given legal efforts to improve the treatment of animals were virtually of equal importance and "always justified." There was also agreement that liberating animals from labs and farms was more justified than causing damage to property where animals were badly treated. The most important finding was that the animal protectionists surveyed in both Australia and the United States believe that legal efforts are more justified than are illegal tactics in improving the welfare of animals, a finding that strongly refutes the "violent extremist" label attributed to activists in the animal movement by some of their opponents.

There was also a strong consensus in respondents' views on the effectiveness of tactics that are designed to improve animal welfare. There was only one minor difference in how the respondents ranked the effectiveness of legal tactics, but there was no variation in the ranking of illegal tactics. On the latter, both American and Australian respondents agreed that liberating animals from labs and farms was more effective than damaging or destroying property where the animals were housed. As with attitudes toward the justification of illegal tactics, the results indicate that legal efforts are seen as more effective than are illegal tactics in improving the treatment of animals in both countries.

However, respondents in both countries agreed that peaceful demonstrations, while always justified, were usually ineffective in getting their animal welfare message across. Thus, while the vast majority of respondents favor peaceful and legal means to achieve improvements in the treatment of animals, many are evidently disgruntled that their conventional lobbying and letter-writing campaigns usually fall on deaf ears. More aggressive, or at least more assertive, action then becomes attractive to those frustrated by peaceful protest activities. While this applies even to the most moderate welfarist, the vast majority of animal protectionists in the United States, the United Kingdom, and Australia draw the line on violence involving harm to fellow humans.

The Experience of Animal Rights Activism: Sacrifice and Satisfaction

Animal protectionists surveyed in ASIS were asked if their involvement in the movement was a personal sacrifice. Respondents divided on the question, with 40 per cent disagreeing and 44 per cent agreeing. Although slightly more see involvement in the movement as a personal sacrifice, nearly 80 per cent of the sample said that their involvement was personally satisfying. What appears to be a contradiction can be explained by the qualitative responses to these questions. For many activists, personal sacrifice on behalf of animals is not seen as a negative cost but rather as personally gratifying. Milly, for example, has invested a good deal of time and money in trips abroad to protest against the live export trade. While this has meant financial sacrifice, it has not been detrimental to her well-being. On the contrary, she describes having met "the most wonderful people," who helped restore her faith in humanity. Many other activists simply said that it was the least they could do given the massive suffering that animals endure.

Likewise, animal advocates saw their work as satisfying rather than as a heavy toll on their personal lives. After all, animal advocates are paid for their trouble, while most grassroots activists have to bear financial and emotional burdens without the support of an organisation. Being an unpaid foot soldier on the front line of the movement is one of the key characteristics of the animal rights activist, and it is one that distinguishes the activist from the advocate. Nonetheless, there were activists like Sherry, the duck rescuer, who derived "job satisfaction" and even a sense of careerism from their involvement as unpaid volunteers in the movement:

I've never had a career. I brought up kids, that was it, my two boys. I was always very shy. I couldn't talk to people. It's done incredible things for me now. I can talk to the media. I do interviews, and I love it with a passion that I've never loved anything before, except my children. Yes, I'm committed to the end, so I believe it is a career . . . I don't want to get a full-time job because I know it's going to take me away from the ducks, and I can't afford to let that happen. So yes, it is a career. (Interview 1994)

Involvement in the protests against live animal exports has meant something similar to Milly, whose day job is as a social worker in London:

I don't see it as a sacrifice because I find it life enriching. I mean, I feel that since I have become actively involved in the animal welfare movement . . . I feel more positive about life. Because I feel that there is nothing more debilitating than feeling that you are a victim, and you are on the receiving end of a lot of ridiculous decisions made by Parliament or by politicians. Once you have reclaimed your own power . . . then it's actually quite liberating in a personal sense as well. (Interview 1996)

Earlier in the interview, she explained what it meant being part of a movement engaged in what she sees as a moral crusade:

I've met lovely people through this movement. It has reenforced my faith in human nature. Up until that point, I was beginning to think that humanity was going down the pan really, that people wouldn't stand up for what they know to be right. . . . But there are still people around prepared to, you know, to put themselves on the line and stand up for what's right. (Interview 1996)

Other activists were less enthusiastic and spoke about the personal cost of battling against *"the brick wall,"* where the only effective means available to breach the wall seemed to be in the form of violent, illegal action. Gaynor referred to activism as *"emotionally draining . . . and at the end of the day when there's no result, it's very demoralising."* Rhett spoke of *"the enormous chasm between me and other people"* outside of the movement. Joan believed that grassroots animal rights activism inevitably took its toll on individuals because *"it also splits up families and splits up communities, as no doubt slavery did."* She articulated a very common experience of animal rights activists:

I think the whole animal rights movement changes everybody's lives, both their family lives, because their families tend then to get a bit neglected, and the attitude towards other people, the attitude towards their lifestyle, you know, from vegetarianism to veganism strains relationships. (Interview 1996)

An episode in Rhett's life illustrates the kind of tensions, albeit much less serious ones, that Joan outlined:

I went out with a woman a few months ago briefly and she said to me, "Would it upset you if I ate meat?" and I said, "No," but it was a lie and she knew it, and that was about the end of it! (Interview 1994)

There are many examples in the data of otherwise like-minded people experiencing similar deprivations quite differently. Phyllis and Joan, for instance, both see their activism over many years as a moral obligation and not in any way as a personal sacrifice. But each experiences deprivations differently. At 76, Joan finds the idea of being arrested unpleasant, but she is nonetheless philosophical about it:

I actually find it quite alarming being arrested, and also it's not very pleasant sitting in the cells, it's a bit claustrophobic. And, of course, if you've been relatively law abiding, particularly me who works in the courts on the other side as it were, I'm glad I have, it gives me experience. But the criminal court's another matter and we've had teach-ins about how to behave in interviews and so on. (Interview 1996)

Her colleague and friend Phyllis, who is more than 10 years her junior, sees the deprivations of a three-day hunger strike as a relatively minor inconvenience compared to what the animals suffer:

We went without food. We did have water, so we were a bit luckier than the animals actually, because they didn't even get water. And having said that, after the end of three days, at least we could celebrate by eating and rejoicing, whereas the animals have their throats cut. (Interview 1996)

We can see from the foregoing discussion that animal rights activism is a complex process involving different identities of *being* and a variety of experiences associated with the *act* of being an activist. These experiences range from being labelled an extremist with the deprivations and risks that it implies to, in the case of most activists, the moral duty of bearing witness for the cause, including the practice of ethical vegetarianism.

The Emotional Toll of Animal Activism

In between legal and more radical illegal tactics, which respectively everyone or no one supported, about a quarter of the ASIS respondents claimed to have taken part in an animal rescue operation, and over half took part in a peaceful protest. The participation in the more risky legal actions turned out to be a good indicator of the different self-descriptors that the respondents used to describe their involvement in the animal movement. Participation in peaceful protest, rather than in the direct action of an animal rescue operation, was the most common form of activism among the ASIS respondents in Australia (Munro, 1996b).

Virtually all activists had participated, at some stage in their activist careers, in a peaceful protest. In the more risky protest involving the rescue of animals, the level of participation declined the most for supporters and the least for activists. The patterns of activism clearly distinguish the activists from the other categories. While most respondents indicated that they did not support liberating animals from labs or farms, a sizeable number had participated in rescuing animals, perhaps in the duck-liberation campaign, which is one of the most high profile conservation/animal rights issues in Australia. While this antipodean antihunting protest is mild in comparison to the more aggressive tactics of the Hunt Saboteurs Association in England, it does have its unexpected dangers and compensations, as Roger explained:

The first bird I rescued nearly took my eye out! . . . The bird just turned around and jabbed straight at me! The sunglasses saved my eye. I remember that bird. I have a deep respect for darters. (Interview 1994)

For many animal protectionists, the attraction of rescuing animals in distress is visceral, and it far outweighs the dangers involved in confronting hunters and those on the other side of animal liberation. Worse than the dangers of confrontation, however, is the emotional toll of activism. For activists like Milly, the favoured campaigning style achieves a balance between the passion of activism and the more restrained style of conventional lobbying:

It, it really, it almost becomes obsessional. It can take over your life. And therefore, for those of us who have other commitments, you have to try your best to pace yourself because what we are in is in a marathon, not a sprint. And it's very easy to get burnt out too quickly too soon, if you try to do too many things too quickly. For me, it has involved demonstrating in Britain, being on picket lines, leafleting, lobbying MPs, lobbying the press, demonstrating in Europe, France and Brussels, contacting your own MPs on the practical and on the theoretical front as well. (Interview 1996)

All of these activities—including several trips to Europe—were at Milly's own expense. Most activists incurred financial costs in the course of their campaigns, but virtually every individual believed that this was the least they could do. Frustration at not being able to do more for animals was a common complaint of activists in all three countries.

Many of the individuals I met during the course of the study sought a midway position between extremism and helplessness. Most did not think of themselves as extremists in any way. For example, the reluctant Australian activist, Al joked about being a self-confessed "ratbag." But neither did many of my informants aspire to bourgeois respectability. Even so, virtually every interviewee and respondent in this study was opposed to violence of any kind and claimed to be interested only in pursuing animal welfare goals via legal means. As in other studies of animal rights supporters (Herzog 1993), most had never engaged in extremist or violent activities. In the ASIS sample, only a dozen out of about four hundred individuals said that they had damaged or destroyed property where animals were confined. This finding was to be expected given the low ranking that ASIS respondents ascribed to illegal tactics. But in the view of one

English activist, and activists elsewhere, frustration often drove peaceful people to extremes. A Quaker of many years standing, Joan explained how *desperate* moderate people were becoming:

Joan: I mean people get sort of desperate, and they shout rude remarks at the truck drivers and also at the police. I haven't seen anybody harming anybody in our movement, but there's a certain amount of what might be called verbals.

L M: Are others like you influenced by Gandhi and nonviolence?

Joan: Well, I would've thought so but some of my language down at the docks, it's not exactly Gandhian let alone Quakerly! I'm not so sure whether it's not counterproductive, but people get so desperate . . . I've never been, I've not been tempted to harm anybody. In fact, all the deaths have been in the animal rights movement. We've had three, haven't we? Two hunt saboteurs and Jill Phipps, and most of the injuries as far as I know have been to protesters. (Interview 1996)

In addition to the deaths and injuries that movement activists suffer, Andrew Tyler finds the denial of animal suffering the hardest thing to bear, along with the stigmatisation of animal protectionists:

As I say, it's not just a question of having to cope with these extraordinary scenes of violence and exploitation that we physically see and read about, but it's the fact that it's denied and that we are mocked, we are called extremists and mad people. (Interview 1996)

While virtually all interviewees were strongly opposed to interpersonal violence as a means to achieving their ends, several believed that the movement was justified in *"going a bit further"* if governments were not prepared to listen to their legitimate grievances. One member of Animal Liberation in Australia, Gaynor, said that she had occasionally been involved in minor acts of vandalism, arguing that damage to property could be justified if it furthered the animal cause and did not result in permanent destruction:

I do a little bit, you know, like with the rodeo, going pulling down the billboards or spraying the "cancelled" stencil over it, so people think it's been cancelled. You know, just little things like that. (Interview 1994)

Gaynor had mentioned the influence of Singer's *Animal Liberation* (1975) in her decision to become more active in the movement. I suggested that in his book, Singer talks a lot about maintaining the moral high ground, not giving in to violent extremism, and I asked if she would agree with that:

Well, yes, I agree with nonviolence and just sort of civil disobedience if you like. The Gandhi method if you like. Sometimes you do have to go a little bit further to get your point across or to attract attention, and that might involve not destruction, but damage to a property or something, but certainly not too extreme and certainly nothing that couldn't be fixed. (Interview 1994)

Someone Threw a Brick Through a Window

"Going a little bit further" was mentioned occasionally by activists, usually in the context of feeling frustrated after all legal means to improve the welfare of animals had failed. In England, the line "Somebody threw a brick through a window" came to symbolise the protesters' growing frustration over governmental unwillingness to stop what they saw as a brutally cruel trade in live animal exports. One of the protesters interviewed at Dover, Phyllis, in her 60s, was asked about the effectiveness of peaceful protest:

To be quite honest, I don't think that just peaceful protests can stop the export of live animals. After all, the thing that actually brought it to the public's attention was the fact that an animal transport lorry, the window got smashed. And that was actually headlines in the paper. Until that happened, it didn't get headlines in the paper. And all the time that we're protesting now at Dover in a peaceful way, it doesn't get into the news. The only time it gets into the news is when somebody breaks the law, which I find is very sad. (Interview 1996).

Milly continued this theme by explaining what she thought motivated extremist action in the animal movement:

Milly: Understandably, people sometimes can get very frustrated and angry and depressed when all legitimate means of demonstrating have been exhausted, and they have met with a brick wall and then people are tempted, say for example to, you know, to do something silly in terms of, you know . . .

L M: Yeah breaking windows?

Milly: Breaking windows or whatever. But at the end of the day, that's usually when they have exhausted every other possibility. (Interview 1996)

This chapter began by distinguishing between animal protectionists in the streets (grassroots activism) and in the suites (organisational advocacy). Animal rights activism and activists come in several guises, ranging from moderate welfarists to the more radical animal rightists and liberationists, as distinct from extremists in the Animal Liberation Front (ALF). Interviewees in the three countries constructed "salient identities" during their recruitment and initiation into the movement. Being an activist in the animal movement was seen as a quest for the middle ground between the fundamentalism of the purists and the pragmatism of the realists. The role of meat or "animal flesh" was identified as a key determinant for designating status (insider versus outsider) and for measuring commitment (animal rights/animal welfare, vegetarian/vegan) inside the movement itself.

Activism in the animal movement is invariably associated in the public mind with the ALF and violent extremism by misanthropes who prefer animals to humans. Evidence from the ASIS sample strongly and unequivocably refutes this as the norm in the animal movement. The labels "compassionate beasts," "creatures of conscience," and "caring sleuths" characterise more accurately the activities and identities of the animal protectionists encountered in this study. The profile that emerges for an animal protectionist in all three case study countries is of an individual who rejects illegal and violent tactics as unjustified and ineffective for achieving improvements in animal welfare. However, when peaceful, legal protests fail, more aggressive, disruptive actions become more acceptable, even to moderates in the movement. The frustration described as *"the brick wall of indifference"* was what drove the thousands of normally peaceful citizens of middle England into the streets during the live export protests in the United Kingdom. What was to become the most important grassroots animal rights protest of the late twentieth century is taken up in the next chapter.

NOTE

1. The American survey by Richards (1990/1992) was completed by 853 subscribers to *The Animals' Agenda*. Because of one or two minor changes that were made to suit the Australian sample (e.g., boycotting was substituted for filing legal suits), the data are not strictly comparable.

Chapter 7

Farm Animals: Media Exposure and Collective Action

> The Tory government won't listen so I have to shout.
>
> —Animal rights activist in Brightlingsea

In terms of numbers, animals raised for food on intensive farms are the most disadvantaged of all animals that humans use. While there are usually strict regulations that protect animals used in biomedical research, this is not the case for their counterparts used for food production on factory farms. Such animals are typically defined as property. Except for transportation and slaughter regulations, conditions for the treatment of farm animals are not included in Animal Welfare Acts; and even here, some animals are more equal than others. Chickens, for example, are excluded from the Humane Slaughter Act (1958) in the United States, which means that the law does not require them to be rendered unconscious before slaughter.

The three grassroots animal protection SMOs in this chapter were chosen because they illustrate how different groups in the case study countries mobilise support for the animal protection cause by focussing on a single issue and making the most of the limited resources available to them. The SMOs are all concerned with the treatment of farm animals, and all originated in the last three decades of the twentieth century: Farm Animal Reform Movement (FARM) in the 1970s in the United States; Mountain Residents for Animal Rights (MRAR) in the 1980s in Australia; and Brightlingsea Against Live Exports (BALE) in the 1990s in England. While all three SMOs share the same ideology of animal protection, they differ in their identities, structures, tactics and, most importantly, in their relations with the media.

A distinction needs to be made between the ideology of a social movement and the identity of a social movement organisation. Most animal protectionists would probably agree that the ideology of the movement—that is, its core beliefs and values—is to protect animals from human exploitation, mistreatment, and cruelty. While each of the SMOs has its own distinctive identity, they share the view that cruelty to animals is an injustice that needs to be addressed. Each SMO has the same diagnosis of the problem—cruelty to farm animals in general (FARM), to farm animals that fast-food chains like McDonald's use (MRAR) and to farm animals that are transported (BALE)—but each has a different prognosis or solution to the problem. Favorable media coverage is usually seen as essential to the success of movements for social change. As I will show shortly, this is by no means a straightforward issue for these three SMOs involved in campaigns on behalf of farm animals.

In common with other new social movements (NSMs)—in the environmental, peace, and women's movements—the three SMOs in their different ways used tactics that were "unorthodox, informal and above all, critical and value-infused" (Crook et al. 1992:155). Crook et al have singled out several characteristics of grassroots activism in NSMs that apply to FARM, MRAR, and BALE:

1. Action is driven by moral rather than by instrumental considerations
2. NSMs are antibureaucratic and rely on self-organisation
3. Key activists are suspicious of conventional party politics and established elites
4. Activists combine leisure activities such as drama and street theatre with protest
5. NSMs are highly dependent on the mass media for getting their messages across to the public (1992:148)

All of these characteristics have been features of animal protection activism during the movement's long history, but it is their reliance on the mass media that is the crucial new ingredient in the quest for animal rights in the latter part of the twentieth century: "The new elements in the contemporary movements are their *specific orientations* combined with international *mass media exposure*" (Crook et al. 1992:148). While Crook and his coauthors see a more symbiotic, mutually dependent relationship between movements and the media, Gamson acknowledges the importance of the media in determining social movement success, but claims that the link is more one-sided: "Much of the impact of any collective action today depends on whether and how it is treated in the media. . . . The movement needs the media far more than the reverse" (1985:618). The three case studies that follow will therefore emphasise the SMOs' relations with the media in their various campaigns.

FARM ANIMAL REFORM MOVEMENT (FARM)

As the name indicates, FARM was originally founded as a reform movement. It actually began as the Vegetarian Information Service in 1976, but changed its name five years later to Farm Animal Task Force following a 1981 conference that FARM cites as the beginning of the animal rights movement in the United States (*The FARM Report* 1996:2). In recent years, it has reverted to its original identity with the mission of "promoting planetary survival through plant-based eating." For a small organisation of only three or four full-time staff members, its campaigns are remarkably successful in attracting wide media coverage and reaching a large audience. FARM's founder, Alex Hershaft, believes that being small allows the organisation greater flexibility to respond to issues and initiate new ideas, something which FARM does very well. For example, it is possibly unique among grassroots animal rights SMOs in that it runs training sessions for activists that are also advertised on their Internet home page.

Red meat and poultry production in the United States amounted to 75 billion pounds in 1996, an estimated consumption rate of 216 pounds per head. Given the size and power of the farm lobby in the land of "Hamburger Heaven," FARM's capacity to make a difference to the well-being of farm animals must be seen as slight. Added to this formidable lobby is the fact that there are no federal laws in the United States that protect farm animals on individual ranches and farms, although some organisations, such as the American Veal Association, have developed voluntary animal welfare guidelines for the care of animals on commercial properties. However, animal protectionists contend that self-regulation is no substitute for the legal protection of animals, and public opinion appears to be on their side. A nationwide poll in 1995 revealed that 93 percent of adults surveyed were opposed to the close confinement of animals like pigs, calves, and laying hens, while 80 per cent believed that the meat and egg industries should be made legally responsible for protecting farm animals from cruelty (Tolles & Dyott 1996). That this is not the case is what motivates the campaigns of groups like FARM, which essentially promotes the boycotting of meat and meat products in the United States.

FARM's best-known campaign is its annual Great American Meatout, which it has sponsored since 1985 to help Americans "kick the meat habit." The idea is to promote vegetarianism as "a more wholesome, less violent diet" by hosting "America's largest annual grassroots dietary education campaign" in all 50 states. The meatout events in the United States have been described as publicity stunts to induce media coverage of meat-eating issues and have resulted in some 200 news stories in the American media, including a front page report in *The Boston Globe*. One newspaper featured a photograph of a campaigner outside a McDonald's fast-food outlet holding placards reading: "Farm animals never have a nice day!"; "McDonald's = McDeath to the rainforests"; and "Try a food revolution—go Vegie!"

Messages of this kind are carried into millions of living rooms in America during the March meatouts each year. Unlike other comparable events, such as Fur-Free Friday and World Day/Week for Lab Animals, the Great American Meatout has been able to avoid a sustained decline in media coverage during the past decade. Dena Jones (1996) suggests that the goals of the animal rights movement have been stigmatised by government and industry as an unacceptable threat to the United States economy, which in part explains the lack of interest by the mainstream media, whose interests are served by supporting the status quo. She notes that since 1990, no single issue, except perhaps abortion, has received significant media coverage in the US. In this context, the coverage of the Meatout must be seen as exceptional. According to Jones, FARM's success is due to its positive approach and the opportunity it provides to collaborate with environmental and vegetarian groups. The view is shared by Scott Williams, the Meatout director at FARM:

> *With veganism or vegetarianism, you're offering people something. In animal rights, you're telling people what they shouldn't do, what they can't do, whereas veganism offers them something in return, which can be a much more positive thing.* (Interview 1996).

Williams describes himself as an animal rights/vegetarian activist who gave up meat five years ago when he saw a truckload of chickens just as he was dining on a hot dog. He remembers the moment precisely—the occasion of his father's fiftieth birthday—when he decided to kick the meat habit. He believes, like Adams (1992), that meat eating is the most extensive and institutionalised form of violence against animals and points out that "if you can eat them, what *can't* you justify?"

FARM is one of five SMOs in the United States devoted exclusively to promoting the rights of farm animals, which they claim account for a massive 96 per cent of all abused animals. The activists at FARM know that their issue is one of the most challenging in the animal rights movement, but they are undaunted by the task. FARM has, in recent years, changed its overall strategy to be more effective. Its previous activist tactics of confrontation—civil disobedience, sit-ins, demonstrations, and the like—have been replaced by education campaigns and advocacy work. Nonetheless, members are encouraged to be active on behalf of the cause, ranging from individuals signing a "no meat pledge" to participating in collective actions, such as street theatre and the like.

Apart from The Great American Meatout, one of FARM's most innovative tactics is the use of media blitzes to get its message across. As noted already, the mass media have not been receptive to single-issue groups in the 1990s, and yet FARM has been able to use the media as a tool for what Lemert (1984) calls "mobilizing information." As the term implies, activists mobilize information by having details of their issues and campaigns "advertised" free of charge in the media. Of course, the media outlets concerned do not do this altruistically. It is the task of the SMO to circumvent the normal checks on what newspapers are prepared to print. In 1994, FARM devised a letter-to-the-editor campaign whereby contacts and members throughout the United States sent "guided" letters to major newspapers, outlining FARM's ideas on a host of issues with an underlying antimeat theme. The contacts were meant to use the sample letter as a guide, rewriting it in their own words and signing it as an authentic opinion. In this way, FARM claims to be able to reach more than 50 million people in the United States, thereby saving millions of dollars in advertising and possibly some animals as well.

However, the *Letters from FARM* project has recently been the subject of a complaint by the National Cattlemen's Association (NCA), which criticised FARM for using false names and addresses on the letters it directed to about 60 newspapers. Hershaft denied that the organisation had been deliberately devious and requested that his members sign the letters they send to the press. Furthermore, FARM insists that they have the public's health interests at heart, while financial interests motivate the NCA (Case 1995).

In any event, the emphasis on health via vegetarianism is considered more appealing to readers than are the moral arguments associated with the animal rights case against factory farms. FARM appears to recognize the limitations of the cruelty approach since its own newsletters have moved in this more positive direction. Kunkel (1995:251) analysed the contents of *The FARM Report* between 1989 and 1994 and found that column inches devoted to health and environment issues have steadily increased over the period; space devoted to cruelty to animals, while still greater than that given to health, was less than the amount of columns for health and the environment combined. In another study that analysed a sample of American newspapers between 1900 and 1976, Kellert and Westervelt (1982) found that moral issues were the least frequently mentioned during the first three-quarters of this century. They concluded that newspapers in the study did not see animal rights and cruelty arguments as newsworthy (Jones 1996).

Kunkel argues that this is the reason behind FARM's "rationale expansion" in its campaign against factory farming: "When certain putative conditions are deeply embedded cultural practices, claimsmakers are unlikely to generate widespread agreement that the condition constitutes a problem. . . . Rationale expansion thus provides an indirect attempt to solve the original problem" (1995:253). In FARM's case then, environmental and health considerations have been added to the original animal rights cruelty frame. While cruelty still features in FARM's documents, references are fewer since these do not resonate with the dominant cultural habit of eating meat in the birthplace of "Hamburger Heaven." People do not want to be reminded of the cruelty involved in their dietary customs, and they are more likely to alter their consumption of meat when given a positive alternative that is both healthy and environmentally friendly. Kunkel notes that it is too early to judge the success of the domain expansion that FARM adopted, but Williams is confident about the general trend towards vegetarianism in the United States. He mentions John Robbins' *Diet for a New America* as the present generation's "lightning rod" for conversion, claiming that the book has been as influential for the present generaton in the United States as Singer's *Animal Liberation* (1975) was for the first generation of activists in the 1970s and 1980s (Interview 1996).

In a special edition of *E magazine* dealing with "the great divide" between environmentalism and animal rights, Robbins explained the need to be more inclusive by seeing people of both persuasions—animal liberationists and environmentalists—as kindred spirits. He rejected the notion of a "great divide" and noted that "essentially the spirit that motivates both of these social movements is one of wanting to move from a dominating, exploitative mentality toward the natural world and its inhabitants to a more respectful one" (Robbins 1995:38). This reflects the philosophy of Mountain Residents for Animal Rights (MRAR), the Australian example of a grassroots animal rights SMO opposing the use of animals for food.

MOUNTAIN RESIDENTS FOR ANIMAL RIGHTS (MRAR)

MRAR was founded in 1984 by Anne Elliott in the Blue Mountains west of Sydney. It has the threefold aim of "saving animals' lives, saving finite resources, and promoting a more gentle way of living" (Elliott 1992:15). These represent the three related goals of the animal rights, environmental, and vegetarian movements that have influenced the activities of MRAR for the past decade.[1] Strictly speaking, MRAR is not a single-issue SMO like FARM since it has been involved in antivivisection, anticircus and antizoo protests during the past decade. But its most important and innovative work has been the "transforming McDonald's" campaign and the related "eco-friendly eating" campaign. In both these campaigns, MRAR has argued that vegetarianism is the best way to protect both animals and the environment, specifically from the rapacious expansion of the fast-food chain McDonald's in the scenic Blue Mountains area. An analysis of these campaigns in the context of this animal rights SMO will show how effective a small group of grassroots activists can be in challenging much more powerful adversaries.

MRAR has used brochures, flyers, posters, articles, letters to the editor, and paid advertisements in the local press to promote their agenda as social justice issues, which the newspaper proprietors found difficult to ignore as they were concerned with local government politics. The McDonald's campaign is important for this reason since the fast-food giant sought local government approval to establish a restaurant against the wishes of a sizeable section of the Blue Mountains community. MRAR asked candidates in the local elections about their position on McDonald's, and the results were duly published in *The Blue Mountains Gazette* and other newspapers. The McDonald's proposal to set up shop in the pristine environment of the Blue Mountains became an election issue. For two years, MRAR and other grassroots campaigners opposed to development in the region were able to hold the fast-food giant at bay.

When McDonald's finally did get council approval to establish themselves in the Blue Mountains, MRAR started its "transforming McDonald's campaign." On the day of the McDonald's opening, MRAR launched a counterattack with "the world's biggest vegie burger" and Regie McVegie, the eco-friendly, non-sexist clown who became the activists' alternative to Ronald McDonald. Like FARM, the Blue Mountains activists made use of computer technology and social movement networks to notify over 200 environmental, animal welfare, and humanitarian groups worldwide about their alternative ideas for McDonald's, which included a vegie burger on every menu. Although the company's response was negative, it did at least respond, suggesting that the campaign was working. And although "the transforming McDonald's" campaign never attracted the publicity that the McLibel trail generated in England (see Vidal 1997), it did manage to give the public a different perspective on the multinational.

For example, several objections to McDonald's have been identified by critics elsewhere, which at some stage were aired free of charge in the local press: (1) environmental degradation through destruction of tropical areas used for raising beef cattle; (2) proliferation of dead-end jobs; (3) high fat and salt content of fast food at McDonald's; (4) unfair competition by forcing local competitors in the food industry out of business (Gunn & Gunn 1991:36). As a commercial organisation, McDonald's is understandably sensitive about criticism of this kind, and its public relations people are quick to defend the company when the dark side of the hamburger industry is revealed. MRAR has avoided attacking McDonald's on these grounds in the manner of the defendants in the McLibel case and has chosen instead to focus on humour and street theatre. In adopting this more lighthearted approach, the activists hope, firstly, to avoid antagonising the litigious company, and secondly, to encourage it to transform itself by adopting a less meaty menu.

The creative use of humour ensured wide media coverage without the more emotional and confrontational tactics that FARM activists used in The Great American Meatout. To be successful, SMOs need to use tactics that resonate with the culture that they are seeking to influence. MRAR has consciously avoided referring to their campaign as a protest and offers instead a positive, albeit value-infused, message via humorous, unorthodox stunts that appeal to the local press and culture. Vegetarianism's health and environmental advantages—when combined with its gustatory appeal—resonate particularly well in a mountain community that sees quality of life issues as more important than a material affluence built on the back of commercial development, often at the expense of the environment.

But the most striking feature of the eco-friendly eating campaign is the free publicity that MRAR got from the local press. Meat eating is for MRAR both an animal rights and environmental issue. As Rissel and Douglas (1993) point out, neither issue is particularly newsworthy to the Australian media. Yet the small contingent of activists in MRAR has succeeded in using their local media for "mobilising information" (Lemert 1984) about their campaigns. Stated bluntly, its fundamental argument is not one that attracts media interest: "With less than ten years in which to turn the tide against the destruction of the earth, the promotion of vegetarianism and vegetarian food is perhaps the single most important environmental and critical issue of all" (Elliott 1992:15). Only the alternative press would be interested in pursuing a story along these lines, as indeed was the case. But the comic figure of the "nonsexist" Regie McVegie and all that s/he stands for is a different matter. Schudson (1989:165) points out how a cultural object—a ritual, a piece of street theatre, or an advertisement—has rhetorical force with either a positive or negative appeal. It seems that Regie McVegie struck the right balance between fun and gravity since there were dozens of photo opportunities featuring the serious message masquerading as a clown.[2]

Both the "transforming McDonald's campaign" and the related "eco-friendly eating campaign" were unconventional, eccentric, and exhibitionist, and therefore newsworthy. In order to maximise media exposure, the content of the campaigns was contextualised and linked to global issues and universal values such as peace, social justice, harmony with nature, and reverence for life. MRAR used the local media "for informing, educating, converting and mobilising" people to their cause (Crook et al. 1992:156). As in the case of FARM's vegetarian rhetoric, it is too early to gauge how successful the MRAR campaigns have been. Vegetarianism does seem to be on the increase in Australia. For example, in the Blue Mountains, more than a score of restaurants now offer vegetarian and vegan fare as a result of MRAR's urging. And since the end of World War 2, there has been a trend away from red meat consumption in Australia and a corresponding increase in the consumption of vegetables (Munro 1993/1994: 19).

Both FARM and MRAR have adopted the strategy of using the media as a resource for "mobilizing information" in their respective campaigns to save animals and the planet via vegetarianism. Like Robbins and Williams, Anne Elliott believes that vegetarianism is the missing bridge in "the great divide" that ostensibly separates animal rights supporters and environmentalists. In linking environmental and animal protection via vegetarianism, they have played down the moral arguments associated with animal rights and instead have promoted the more palatable message of eco-friendly, healthy eating with the use of colorful events that appeal to the media. FARM's Great American Meatout and the image of Regie McVegie launching "a Big Mac attack" provide entertaining spectacles for the media, either as graphic photographs or television images. Both SMOs then succeed in getting their issues exposed to a large audience on a scale well beyond the resources of small grassroots activist groups.

Yet for agents of social change, media-SMO relations are fraught with difficulties. Several writers, most notably Herman and Chomsky warn that the mass media are nothing more than propaganda machines that "manufacture consent" on behalf of ruling elites (1988). Tarrow is similarly pessimistic when he asserts that social movements possess little cultural power when pitted against "the inherent power of the media to shape perceptions" (1994:23). A more optimistic argument is put by Gamson (1992:71), who recognises the central importance of media discourse in framing issues for the public, but warns that media practices can both help and hurt a social movement's efforts to communicate its issues to an attentive public. Rootes succinctly explains the dilemma for activists: "A movement may seek to exploit the media's insatiable appetite for novelty and spectacle, but no movement without a very broad social base and very considerable resources of power can hope to dictate the terms of the transaction or its outcome" (1984:6). Of the three animal rights events studied by Jones (1996), the Great American Meatout was the least dependent on media coverage. According to Jones, even without media attention, the Meatout introduces thousands of people to a meatless diet through the many community events held throughout the United States (1996:21). Even so, media coverage of these events—claimed by FARM to reach fifty million Americans—is necessary if vegetarianism is to become the dietary future that activists like Williams and Elliott envision.

BRIGHTLINGSEA AGAINST LIVE EXPORTS (BALE)

In early January 1995, there were newspaper headlines and front-page coverage of what the British media initially dubbed a "bunny-huggers' battle" (Ridley 1995). Compassion in World Farming (CIWF) had been campaigning for about two months at Shoreham in West Sussex against live animal exports when their peaceful protest was, according to CIWF, "hijacked by rent-a-mob." The peaceful scenes were contrasted with *The Guardian*'s photograph of a militant animal rights activist smashing the lights of a lorry carrying the animals to the port. CIWF director Joyce D'Silva was reported as having distanced the moderate animal welfare organisation from the "extremists and hooligans who had abused her staff" and was said to be appalled by the violence that attracted widespread media interest. D'Silva denounced the illegal actions, which she said her organisation of 10,000 members would never condone. The live animal export protest was, from the outset, framed by the media as a law-and-order story, and not as a protest on behalf of animals. This chapter explores how this frame came to dominate the news and what the consequences were for the animal movement.

Sources used in this case study consisted mainly of interviews with members of BALE and local residents in Brightlingsea, including town council employees and other eyewitnesses, such as one of the journalists who covered the protest from the outset. I visited the town in August 1996 when the protests had shifted to Dover, the last remaining port for the live export trade. The original founders of BALE continued to hold weekly meetings at the community hall to keep people informed about animal welfare issues in general. It was clear from the meeting I attended that the approximately 60 people there had been politicised by events in Brightlingsea during the first ten months of 1995. By meeting every week, the community of "morally motivated comrades" (McAdam 1988) demonstrated their need to sustain the collective sense of empowerment that they had experienced during the battle for Brightlingsea. Referring to McAdam's study of Freedom Summer volunteers, Lichterman's description applies equally well to the Brightlingsea protesters: "Mass protests and extraordinary resocialization experiences felt good and produced a new sense of self" (1996:188).

Background to the Protest

"The Tory government won't listen so I have to shout." This was the message worn by a protester who was featured on the cover of *New Statesman & Society* (27 January 1995a) in the magazine's story on the nature of protest entitled "England Rising." Brightlingsea Against Live Exports (BALE) was one of a number of SMOs that were formed to protest the mistreatment of animals during transportation from British ports to overseas destinations. BALE's foundation president was Tony Banks, MP, a sitting Labour Party parliamentarian. This grassroots group, like dozens of others throughout England, was set up to give the townspeople a single voice in their campaign against live exports. BALE attracted more media coverage than did other similar groups because of the large number of local residents who supported its protest. At one point, two thousand five hundred residents confronted the police in Brightlingsea, thereby stretching the resources of the authorities' power to protect the agricultural industry and guarantee the future of British trade.

So effective were the Brightlingsea protests that the Tory government invoked the 1986 Public Order Act, which had the effect of BALE disbanding itself temporarily for fear of prosecution. Evidently the government saw the protest as a threat to public order and an intolerable financial burden on the state because of the loss of export income and the heavy cost of policing the daily protests. This had the effect of broadening the issue for the protesters, many of whom were politicised by these events.

For most of the Brightlingsea protesters, the state's use of force to suppress them was a serious breach of trust since they claimed their right to demonstrate was not only a British tradition, but also in accordance with the feeling of the town. And in fact, a survey that the local council commissioned concluded that more than three-quarters of the residents were against the use of their port for live export trade (Tanenbaum 1995). As the representatives of the state, the police force became the subject of many serious complaints by the protesters, some as late as October 1995, eight months after the demonstrations began:

> The past weeks have been extremely turbulent and many protesters have sustained severe injuries, some requiring hospital treatment. These injuries were sustained whilst lawfully demonstrating. *(BALE's Newsletter* 1995:4)

Several people I spoke to said that they had never previously been involved in protests but had become quite radical after seeing the lorries with their pitiful cargo literally pass by their front doors. Many were incensed by the conduct of the riot police and were openly hostile toward the government's handling of the protest. I met at least two people—a man in his early 30s and a woman in her early 40s—who claimed that overzealous police officers assaulted and injured them. Other residents told me that the police no longer had the confidence of the townspeople and that even the children had lost respect for the law, which one person claimed was the reason for a sudden increase in youth crime in the town. There is no doubt that the events of early 1995, which put Brightlingsea on the front pages of British newspapers, also radicalised many of its citizens.

Media Mania: The Battle of Brightlingsea

Benton and Redfearn's (1996) postmortem of the Brightlingsea campaign suggests that animal rights was not the primary issue, although my informants in BALE, as their name implies, claim that it was. And for most members, animal welfare, in contrast to animal rights, continues to be the group's *raison d'etre*, although the wider issue of social justice appears to have been embraced by some of the group's members after their encounters with the police. Two one-hour videos called "The Battle of Brightlingsea," dedicated to the memory of Jill Phipps (who was killed during the protests at Coventry airport) record the turbulent events of 1995. And at least one photographic exhibition has been held to commemorate the town's stand against live exports. But it was the mass media that gave the issue its national and international prominence.

English ports used in the live export trade became the scene for some of the biggest demonstrations seen in the United Kingdom since the miners' strike in the mid-1980s. The year 1995 began with British newspapers trumpeting a moral panic, with headlines about "animal rights siege," "single issue hooligans," and "bunny-huggers do battle." An editorial in *The Times*, headed "Cuddly Terrorism," described the animal liberation protesters as "on a par with the IRA" (*The Times* 8 February 1995), a claim that was often repeated in the media during the mass protests that year. To be sure, other themes were also featured in the mainstream press, although the law-and-order story was the predominant frame for most of the time. Toward the end of the protest, one regional newspaper at least seemed to have a grudging respect for the activists.[3]

Brightlingsea was the scene of protests against live exports from 16 January to 27 October. For most of 1995, BALE was engaged in a public relations battle to combat the negative media framing of their campaign. According to Benton and Redfearn (1996:55), there were three main complaints that the activists had against the mass media: showing the police in a positive light, stereotyping the protesters as "punky types," and distorting the issues. A close examination of the newspaper coverage of the protest shows that the activists' injustice frame never seriously competed with the media's legitimating frame.

The early protests against the live animal trade focussed on the small Essex port of Shoreham, where *The Independent's* headline described the protest, not as an animal welfare issue, but rather as a law-and-order issue: "1,000 police drafted to animal port." The headline was accompanied by a picture showing a hooded animal rights protester smashing a lorry window with a brick (*The Independent* 5 January 1995). The text read, in part:

> They leapt on to lorries, smashing lights and hurling stones through windscreens. . . . West Sussex is a hot-bed of animal rights groups. . . . The Animal Liberation Front and Justice Department, two extremist groups, have a strong local base, which could be used to paralyse the port. But the vast majority of animal rights' groups oppose the trade peacefully. (Penman 1995:2)

The local press framed the protest in a similar fashion. *The East Anglian Daily Times* of 9 January 1995, for example, framed the story in terms of the financial cost that the protesters imposed on the community and highlighted the violent clashes between animal rights extremists and police. The reporter noted that 1,000 police had arrested twenty one protesters who had unsuccessfully tried to stop the lorries of calves and sheep from boarding the ferry to France. When the protesters did eventually succeed in closing Shoreham, and later Plymouth, the lorries were directed to the tiny port of Brightlingsea in East Anglia, about one hour from London.

Most of Brightlingsea's population of 7,000 people became involved in the protests over the next ten months, during which the town's placid existence changed overnight. *The Independent*'s story, "Sleepy port that joined the front line in trade war" (Cusick 1995), summed up the initial reaction of the townspeople as being shocked by the Battle of Brightlingsea, which saw 500 protesters, consisting mainly of locals with a smattering of hunt saboteurs and anti-road protesters, block the only road that leads to the wharf. On the following day, *The Independent* ran a full-page, illustrated story about the protest that showed large numbers of local residents in confrontation with police in riot gear.

What had been a law-and-order issue briefly turned into a police brutality story, which presented the police as overreacting to what was essentially a peaceful protest: "Children and elderly people were caught up in violent scenes," according to *The Essex County Standard* of 20 January 1995, and "many were left in tears after being dragged away by police." Needless to say, there were some graphic images, including one of an 85-year-old woman admonishing a large policeman in riot gear. The story headline, "A tyrannical penalty against a small town," referred to allegations of police brutality in the arrest of 21 residents (*The Independent* 20 January 1995:5). Yet there was still no real attempt to frame the story from the protesters' perspective, that is, as a protest against cruelty to farm animals. Some of the more upmarket newspapers and magazines were at least taking a more analytical stance toward the sheep wars when they focussed on the protesters themselves.

Protesters as "bunny-huggers," "crazed spinsters," and "urban terrorists"

At the outset of the "Battle of Brightlingsea," *The Observer* (Eagar 1995:3) noted that what was once a middle-class armchair preoccupation in animal welfare had become "one of the most explosive issues in British politics," while *The Economist* saw the protest as unifying a hitherto fragmented movement:

> The issue of animal exports has united under a single banner the animal-welfare movement—from the moderate, mass membership Royal Society for the Prevention of Cruelty to Animals (RSPCA) to the loopy and violent Animal Rights Militia. (1995b:62)

A more sympathetic, pro-demonstrator stance was taken by Paul Binding in *The New Statesman and Society*, who believed that the protests had profound cultural roots and represented "a thorough-going compassionate identification with sentient beings other than the human. It is a movement that affects one's attitude to just about everything" (Binding 1995:17). It was hardly surprising that the passion of the protesters should arouse such sentiments. But what surprised most people in the United Kingdom was the social composition of the people who made up the opposition to the live animal trade. While there had been the inevitable references in the press to "rent-a-mob" (Erlichman 1995), the pictures revealed that many of the protesters were respectable, middle-class men and women. Many told reporters that they were Tory supporters and first-time demonstrators. Mrs. Ferres-Gui, age fifty-five, for example, was described as a businesswoman and former Tory councillor who reportedly usually saw demonstrations as "beyond the pale," but like many others in Middle England, she had been "converted" by the horrors of the live export trade. She explained:

> The first night I turned up I was expecting yobbos and roughnecks. It wasn't like that at all. I saw some students who were peculiarly dressed, but the majority were just ordinary people who care about animals. There were several ladies in their late 80s. (Marks 1995:11)

Undoubtedly the sight of "Middle England" protesting in the streets severely shook the conservative British government. And the fact that the protests continued despite massive state intervention underscored the reputation of the British public as a nation of animal lovers. It also ensured the continued interest of the electronic media. As Chomsky (Achbar 1994:165) observed, the one-off protest is of little political consequence, but authorities are genuinely flustered when protesters continue to return to the streets.

However, there were some observers, like Auberon Waugh, who were not impressed by the animal rights demonstrations. Writing in the conservative magazine *The Spectator*, Waugh dismissed the protesters as "crazed spinsters" and praised the police as "the thin blue line between society and all the forces of malevolence and depravity seeking to destroy it." The police, he thought, were "usefully employed against a very real threat to the nation's agricultural industry" (Waugh 1995:7) Waugh's "crazed spinster" label was reminiscent of the attacks made on female antivivisectionists a century earlier, which were described in Chapter 1.

Others saw the protesters as even more sinister and claimed that their activities were acts of terrorism: "The greatest terrorist threat in the country at present probably comes from the Animal Liberation Front, the extreme arm of a movement that is largely middle-class and female" (Ridley 1995). It is clear from the article that Ridley equates the ALF with the "urban bunny-huggers," whom he identifies in the person of former model Celia Hammond, one of the "middle-class, articulate women confronting riot police on behalf of calves in of all places Shoreham, Sussex." Damaging property is for Ridley—a dairy farmer in Northumberland—a greater sin than is the cruelty to animals that the terrorist/bunny-huggers want to end. He quotes Hammond's position disapprovingly: "If a windscreen needs to be smashed to bring this to public attention, then I'm behind it." Time and again, it was this act of *throwing a brick through a window* that came to symbolise, for the media, the protesters' campaign against live exports. It was the image that served to construct the typical protester as an urban terrorist rather than as a harmless bunny-hugger/animal lover.

Animal rights fanaticism became an explicit media theme after the accidental death of one of the protesters, Jill Phipps, age 31, who died in early February when she tried to block a lorry load of calves of its way to Coventry.[4] This death inspired several members of the commentariat to offer explanations of this protest for which some people were prepared to die. For example, in arguing against the protesters, Alasdair Palmer suggested that the animal movement had become "a refuge not just for misanthropes and misfits, but also for do-gooders" (Palmer, 1995:16). This argument was evidently supported by Charles Moore, who claimed that the demonstrators had found "a new group of creatures who would never answer back" (Moore 1995:7). An article, headed predictably "Why the elderly are revolting" (Neale 1995:21), canvassed opinion from the public and private sectors that suggested the reason had to do with a backlash from old people who were fed up with being marginalised.

In all of these reports there were few hints, if any, of the protesters' anger concerning the cruel treatment of animals in the live export trade. Indeed, a cursory examination of the newspaper stories during the height of the protest offers little indication that the protest was about animal welfare at all. The headlines tended to frame the story not in terms of a campaign against cruelty to animals, but rather as a narrative of protest[5] in which the chief elements were "protest," "clashes with police," "confrontation with authority," "animal rights siege," "frontline battles," "defeats and victories," and "militancy." In the main newspapers, the issue of animal welfare was featured in less than one in five articles in *The Independent* and in only one in nine cases in *The Times*. While it is true that most of the pictures reflected the law-and-order theme of the coverage—with graphic photographs of police in confrontation with protesters—there were several potent images of animals in apparent distress that undoubtedly served as "prods to action" (Snow & Benford 1988) for many readers. Several of my interviewees in the United Kingdom cited television images of animals in the transport lorries as the initial stimulus for their participation in the protests against live animal exports.

WIDER IMPLICATIONS OF THE CAMPAIGNS FOR FARM ANIMALS

One activist, James, believed that the protests against live exports represented an historical turning point for the animal rights movement in England:

> *I can't think of any issue in any movement, or any particular issue regarding animals or animal welfare in the world, that has ever had such support compared to here in the UK with live exports. And I think, I mean eventually this must win. It's not old, live exports. It's a very recent thing and it is winnable.* (Interview 1996)

The head of research and communications at CIWF, Tim O'Brien, was more circumspect. He agreed that the media interest in the protests in England was intense and that the live export issue "opened the door" to other important animal issues. However, he likened the "explosion of interest" in 1995 to what happened in the green movement in the mid-1980s, and he was fearful that the interest would soon dissipate:

> The intensity of public interest in the welfare of farm animals has been much greater in BALE's protest than in the campaigns run by FARM in the US and MRAR in Australia. In the English case study, unlike in the American and Australian examples, there was massive intervention by the state to quell the protests. When the Public Order Act of 1986 was invoked in Brightlingsea for example, BALE disbanded out of fear of prosecution and 350, rather than 2500 people went on to the streets. (*New Statesman and Society* 21 April 1995b:5)

What distinguished BALE from the FARM and MRAR campaigns on behalf of farm animals was the threat it posed to the state. More than in any animal rights campaign in recent history, the British protests against live exports showed, as Rochon argues, that the power of a social movement lies in its militancy, size, and novelty (1990). These features were bound to attract the media, which generally portrayed the activists' protests on behalf of farm animals as the ravings of eccentrics or as matching the IRA in their violence and deviance. Under these circumstances, state intervention was inevitable since the protesters constituted a potential threat to the agriculture and farming lobbies in England.

Governments might be willing to let FARM and MRAR activists run eco-friendly eating campaigns, or even "transforming McDonald's campaigns"—as radical as the long-term effects of these campaigns potentially are—but are unwilling to accept either the massive losses in export earnings or the heavy burden of policing street protests. Early in the campaign, Parliament estimated the cost of policing the protests to be £6 million in England and £2 million in Brightlingsea alone (*New Statesman & Society* 21 April 1995b:5), while the live animal trade to Europe was estimated to be worth £200 million in 1993 (Erlichman 1995:6). Opposition to live exports in the United Kingdom was not therefore to be taken lightly by the state, for as *The Economist* observed, animal rights had emerged as Britain's closest thing to a mass social movement (*The Economist* 19 August 1995c:17).

It was also an issue that inspired many of the animal protectionists in the three countries in this study. For example, ANZFAS in Australia ran an interview with CIWF director D'Silva in their magazine, which also included comments from actress Joanna Lumley on the evils of the trade (*Animals Today* May-July 1995). Activists in England were well aware of Australia's involvement in live exports, just as MRAR was familiar with the tactics that FARM used in the Great American Meatout. In short, animal protectionists worldwide maintained links through the Internet, and more commonly, by exchanging newsletters and magazines.

The campaign against live exports in the United Kingdom was, for a short time in 1995, the single most important issue in the animal rights world. For many animal protectionists—activists, advocates, and supporters alike—it reinvigorated the movement by providing much needed publicity for the animal cause. The fact that this was usually unfavorable did not seem to matter to the movement, which many people thought had reached its nadir. For the professional animal advocates in the suites, the media's interest in animal rights activism was a much needed boost to their campaigns. In the remaining chapters, I look at some of the most important of these in the three case study countries.

NOTES

1. The material in this chapter on MRAR is drawn from Munro's "Grassroots Animal Rights Activism: The Case of Mountain Residents for Animal Rights" in *Journal of Regional Issues* 1994:28, 37–48. The article contains a more comprehensive account of the group's three main campaigns and was based on conversations with MRAR's founder, Anne Elliott, who generously gave me access to the group's files.

2. Local newspapers printed headlines such as *Blue Mountains Gazette* "Regie pushes for eco-friendly burger" (*Blue Mountains Gazette* 8 September 1993), and "Regie McVegie for Major" (*Blue Mountains Whisper* October 1990). During the three-year campaign, MRAR produced a *McDonald's Campaign Pack* that was sent to dozens of media outlets, many of which gave it some coverage.

3. In the lead up to the hunger strike in London, the *Cambridge Evening News* (22 July 1996) wrote:

> The usual Cambridge cranks will be among the loonies in a hunger strike next week. . . . True, history will eventually recognise these cranks and loonies as heroes in the long struggle against cruelty to exported farm animals. History will see their dotty little gesture outside the Ministry of Agriculture as one of the few significant steps towards real civilisation in an otherwise benighted age.

Activists from ARC proudly displayed this clipping on their noticeboard at their regular meetings and at the Saturday stall that I attended.

4. A sample of headlines illustrates the tone of the coverage:

"Veal protester killed as she leaps at lorry" (*The Times* 2 February 1995)

"Making of a veal martyr" (*The Weekly Telegraph* 5–14 February 1995)

"Mother ready to die for animals" (*The Guardian Weekly* 12 February 1995)

"Veal lorry victim had record of militant protest" (*The Times* 3 February 1995), which was preceded by a statement attributed to the dead woman's boyfriend: "Jill would almost be happy to sacrifice her life to halt export of calves from Coventry."

5. In my analysis of another animal liberation campaign, I describe how the images of suffering animals helped the cause of animal welfare more than the media probably intended. Although the language of the television reports highlighted the "deviance" of the "duck liberationists," the protesters saw the accompanying footage as favourable to their campaign (Munro 1997b).

PART III

ORGANISATIONAL ADVOCACY IN THE SUITES

In Part III, some of the most important campaigns sponsored by the organisational wing of the animal movement are described. Chapter 8 begins by referring to the long history of animal advocacy, which included the European custom of putting animals on trial for an assortment of alleged misdemeanors. Modern animal protectionists, on the other hand—including the movement leaders profiled in the United States, the United Kingdom and Australia—seek to punish animal abusers for various offences against animals; the chapter lists a dozen such offences that animal liberationists in the case study countries want to outlaw. One of the most recalcitrant issues for the movement—the widespread practise of animal experimentation—is the focus of Chapter 9. The views of a range of animal protectors are reported, from moderates and pragmatists to abolitionists, who see vivisection as an unmitigated evil that remains the animal movement's greatest challenge. Cruelty to domestic animals, particularly dogs, is discussed in Chapter 10; the chapter reveals that despite the generally strong ideological consensus on what constitutes cruelty, movement insiders hold conflicting views about the role of companion animals in modern society. Even so, large and small animal advocacy organisations use the emotional appeal of pet animals to promote the broad cause of animal welfare. The campaigning strategies of direct action and fund-raising via direct mail are analysed in the context of the movement's efforts to mobilise compassion for maltreated animals in the West and in less-developed countries, especially captive animals such as bears used for entertainment. For many animal protectionists, the campaign against recreational hunting in the West has the most promise for success. Chapter 11 discusses antihunting campaigns in the case study countries and profiles some of the movement's most high-profile leaders in this particular quest for animal rights.

Chapter 8

Multi-Issue Advocacy in the United States, the United Kingdom, and Australia

> We have to shock and mesmerise and entice, and tell powerful stories about the suffering of animals.
>
> —Andrew Tyler, Animal Aid

New social movements engage in "a form of advocacy which challenges the 'reality' of dominant values. It is a process of making an alternative form of knowledge count" (Harries-Jones 1991:5). This chapter focusses on the way that animal protection social movement organisations (SMOs) in the United States, the United Kingdom, and Australia challenge existing ideas and attitudes about animals by using their resources to advocate the rights of animals. Animal advocacy has a long and sometimes curious history, particularly in Europe where the earliest legal defences of animals charged with wrongdoing were conducted in the secular courts. One of the strangest sources in the history of animal-human relations is E. P. Evans's *The Criminal Prosecution and Capital Punishment of Animals*, which described animal trials from medieval times up until the last recorded case in 1906, when the book was published. Evans documents the medieval practice of bringing recalcitrant animals and insects to court for alleged crimes such as homicide and theft. Dolphins, dogs, pigs, and even locusts, flies, bees, field-mice, termites, weevils, and woodworms were among those prosecuted—usually in absentia—by the court, which provided the accused animals with their own defence counsel, who were certainly the first official animal advocates in history. One of the most famous *plaidoyers* was Bartholome Chassenee (born 1480), who in one trial defended his clients—a family of woodworms—against the charge of wilful destruction to church property in the village of Mamirolle, France. The excerpts from Chassenee's defense may appear ludicrous to the modern reader given the oddness of a trial involving absent woodworms, but his arguments in defence of animals' rights strike a more familiar note:

> First, my clients, though notoriously dumb, have been summoned to explain their behaviour . . . [The summons] is not [lawful] since my clients are brute beasts acting only from instinct . . . What have these humble *bestioles* done since the day of Creation but exercise the inalienable rights conferred on them at that time, rights which Man has no power to curtail or abrogate? . . . [E]xcommunication . . . is both inappropriate and unlawful [for] how can this be an appropriate punishment for a *bestiole* which does not possess an immortal soul? I ask, therefore . . . that the defendants be acquitted and exempted from further prosecution. (Barnes 1989:65–69)

What is interesting about this testimony is the seriousness with which the proceedings were conducted. Secular courts in Europe, right up to early in the twentieth century, afforded animals the same rights as humans to receive the due process of law. We can detect in Chassenee's full testimony the earliest notions of animal rights that today's more sophisticated advocates invoke on behalf of their "clients." There are also some obvious differences. For instance, the claim that animals are mere brutes has recently been challenged by the Reverend Dr. Andrew Linzey, who was appointed to the world's first academic chair in ethical and theological aspects of animal welfare at Mansfield College, Oxford. According to Linzey, the International Fund for Animal Welfare (IFAW) provided the college with a grant of £250,000 over five years to investigate a dozen questions concerning theological issues in animal welfare. Linzey argues that Christian beliefs are not inherently detrimental to animals and suggests that what moral status animals enjoy today can largely be attributed to a Christian counteroffensive that sees cruelty to animals as nothing less than intrinsically evil.

A sense of his strong feelings about animals can be gauged from the chapter headings in the second part of his book, *Animal Theology* (1994): "Animal Experiments as Ungodly Practice," "Hunting as the Anti-Gospel of Predation," "Vegetarianism as a Biblical Ideal," and "Genetic Engineering as Animal Slavery." Elsewhere, he argues that animal advocates need to go beyond rational thought in arguing the case for animals since animal advocacy depends on "deeper capacities for imagination, insight, creativity and so on" (*Animals Today* November–January 1994–1995:16). Many of these ideas may be as strange to the modern reader as those in Chassenee's testimony, but minus their Christian overtones, they are not far removed from the conventional arguments of modern-day advocates in the suites.

Table 8.1
Mean Scores of Australian and American Activists' Attitudes Toward the Use of Animals

	Mean Rating[a]	
	ANZFAS Australia	*Richards USA*
1 Using steel-jawed leghold traps to capture wild animals	1.02	1.06
2 Using animals in cosmetic and beauty product experiments	1.05	1.13
3 Killing an animal to make a fur coat	1.16	1.17
4 Selling unclaimed dogs from animal shelters for use in medical experiments	1.19	1.29
5 Hunting wild animals with guns	1.32	1.49
6 Exposing an animal to a disease as part of a medical experiment	1.33	1.62
7 Raising cattle for food in feedlots	1.34	1.75
8 Using horses for jump/steeple racing	1.79	2.68
9 Eating meat	2.81	2.74
10 Keeping animals in zoos	3.08	3.02
11 Raising cattle for food on open range or pastures	3.48	3.31
12 Killing rats in residential area	4.93	4.24
13 Killing cockroaches in a residential area	5.35	5.34
14 Keeping a dog or cat as a pet	6.31	6.49
15 De-sexing a pet	6.64	6.62
	Mean 2.85 (SD2.03)	Mean 2.93 (SD1.92)

Note: [a] Rating scale values range from 1 (extremely wrong) to 7 (not at all wrong). SD designates standard deviation.

ISSUES FOR ANIMAL PROTECTION SOCIAL MOVEMENT ORGANISATIONS

All four practices that Linzey identifies—experimenting on animals, hunting, meat eating, and more recently, genetic engineering—are among the principal targets of animal protection SMOs in the three countries in this study. It was clear from the interviews with the spokespersons in the SMOs as well as from their organisations' promotional literature that these are among the most important issues for the animal rights movement worldwide. Some idea of the range of animal welfare concerns, and the importance given to these particular issues, can be seen in the responses of Australian and American respondents to the Animal and Social Issues Survey (ASIS) in Table 8.1.[1] The 15 issues consisted of about two items for each of the movement's main concerns, namely vivisection, hunting and trapping, agribusiness, sport and entertainment, companion animals, and killing animals for their products or as pests.

Hunting and trapping, as well as experiments that use animals for product testing or for medical reasons, rank high on the list, while meat eating does not. What is most remarkable about this cross-national comparison of attitudes is the identical ranking of the respondents, which suggests an unexpectedly high degree of ideological consensus among animal rights supporters in Australia and the United States. As only a small number of people in the United Kingdom completed the survey, it is not possible to say if the consensus applies to them as well, although the limited data suggest that there are more similarities than differences.

With the exception of items twelve and thirteen (killing rats and cockroaches in residential areas), all of the practices in Table 8.1 are the concern of one or another animal protection organisation in Australia, the United Kingdom and the United States. In what follows, I describe some of the main features and preoccupations of what can be described as national multi-issue animal advocacy organisations: the Australian and New Zealand Federation on Animal Societies (ANZFAS)/Animals Australia, Animal Aid in the United Kingdom, and the Animal Welfare Institute (AWI) in the United States.

MULTI-ISSUE ANIMAL ADVOCACY SMOs: THREE COUNTRY CASE STUDIES

These particular SMOs were chosen first because they are examples of national multi-issue SMOs that broadly represent the three main approaches to animal protection—animal welfare (AWI), animal liberation (ANZFAS), and animal rights (Animal Aid)—and second because they are exemplary animal advocacy organisations that exhibit features that may be unique in the animal movement: ANZFAS is arguably the only federation of regional animal societies in the world; Animal Aid is unusual among animal rights organisations in that it promotes a philosophy of radical veganism; and the Animal Welfare Institute in Washington D.C. bases advocacy largely on its intellectual work, principally the publication and dissemination of quality scholarly books, monographs, and articles.

In their various campaigns, all three SMOs use what Eyerman and Jamison (1989) call "knowledge as an organisational weapon." In their study of Greenpeace, they argue that this quintessential environmental SMO "can be defined primarily by its transformation of knowledge into an organisational weapon" (1989:113). They emphasise the crucial importance to Greenpeace of disseminating selective environmental knowledge as "strategic information for the people" (1989:114). According to Eyerman and Jamison (1989), it is the selective use of environmental knowledge as an organisational weapon that gives Greenpeace its advantage in public fora. They suggest that because the organisation is success-oriented rather than value-oriented, it is more interested in the dissemination of attention-catching information than in endless debates about philosophy. Direct mail, the production of its own video and film archives, and conventional advertising are characterised by a professionalism that ensures that the environmental message is kept simple and will resonate with as wide an audience as possible. Yet as anyone who has received Greenpeace literature in the mail knows, the message is typically emotional rather than intellectual.

In an evaluation of Greenpeace USA, Shaiko (1993) points out that more than any other environmental organization in the United States, Greenpeace consistently markets itself in starkly emotional terms. It attracts people by the colourful, eye-catching packaging of its campaigns and its invitation to participate vicariously in dangerous, albeit nonviolent, direct-action stunts designed to save charismatic wildlife from the rapacious appetites of whalers and sealers. Shaiko's analysis offers a corrective to Eyerman and Jamison's European-based survey by suggesting that Greenpeace USA employs the basic strategy of guerrilla theater, or ecodrama, to promote its campaigns rather than to promote knowledge per se. Eyerman and Jamison are aware of the tensions that exist between the success and value orientations in the organisation, but they gloss over these in referring to what they describe as Greenpeace's "general cognitive orientation" (1989:103). In its campaigns to defend wildlife, Greenpeace is no different than high-profile animal rights SMOs in exploiting anthropomorphism and in using more objective information and knowledge to win support for their cause. By definition, grassroots activists are oriented toward colourful, ecodrama-style protests, while their counterparts in the suites use the more conventional forms of advocacy to achieve the same ends with different audiences.

In the three multi-issue SMOs discussed below, knowledge appears to count more than emotion does. For ANZFAS/Animals Australia and the AWI, lobbying their respective governments is the principal strategy and hence, for these organisations at least, knowledge—in the form of detailed written submissions to official inquiries and in the provision of expert witnesses or members of ethics committees—is the main tool for accomplishing their goals. Animal Aid in the United Kingdom sees itself as more uninhibited and embraces both a grassroots epistemology and the conventional cognitive style of organisational advocacy.

Thus, the three animal protection organisations employ knowledge in different ways. Singer's presidency of ANZFAS provides the organisation with intellectual credibility since his name is synonomous with animal liberation philosophy; its director, Glenys Oogjes, sees herself as a conduit for empowering others with information. Animal Aid places great store on their intelligent, verifiable packages of information that they use to tell powerful stories aimed at people's hearts and minds. Finally, the AWI's intellectual work consists of producing high-quality, in-house print and electronic media that can be used in their campaigns on behalf of animals; the AWI, more than ANZFAS/Animals Australia and, to a lesser extent, Animal Aid, uses its own in-house publications and electronic media as a distinctive organisational tool.

ANIMALS AUSTRALIA/ANZFAS

The Australian and New Zealand Federation of Animal Societies (ANZFAS), which was discussed in Chapter 5, changed its name in October 1996 to Animals Australia to distinguish itself from the umbrella organisation and its affiliation with New Zealand. Animals Australia is one of the largest animal welfare organisations in Australia, and it is the umbrella organisation for about 35 animal societies in the country. I use the name ANZFAS when the activities of the organisation encompasses both countries, and I refer to Animals Australia when only Australian activities are concerned. The Australian contingent, Animals Australia, numbers approximately 33,000 members compared to about 7,000 in New Zealand.

The federation of regional societies in New Zealand and Australia is a singular achievement for countries with a land mass the size of the United States and United Kingdom combined, and it is one that has the admiration of movement leaders elsewhere (e.g., an editorial in *The Animals' Agenda* 1989a). Moreover, the societies cover the broad spectrum of animal protectionists, including welfarists in the RNZSPCA, pragmatists in the various state branches of Animal Liberation, abolitionists in Animal Rights, other antivivisectionist groups in both countries, and conservationists and environmentalists in groups such as the Kangaroo Protection Co-operative. Although almost two-thirds of its private members claim to be members of the RSPCA, Animals Australia is animal liberationist in orientation, in keeping with the activist, pragmatic style of its president, Singer.

Animals Australia favours an advocacy style of institutional lobbying combined with the grassroots activism of some of its member societies. It eschews violence of any kind and believes that its strategy of institutional lobbying is the only feasible one for an organisation that must negotiate with the Australian government on animal welfare issues. ANZFAS sees its main strengths in research and communication and cites the establishment of animal welfare advisory committees in Australia and New Zealand as confirmation of its success in changing public attitudes toward animals.

Glenys Oogjes: Pragmatic Animal Advocate in the Suites

The director of ANZFAS Glenys Oogjes, has been involved with animal protection since the early 1980s and has been the full-time director of the organisation since 1983. She explains that she first learned of animal exploitation—specifically, the use of animals for testing cosmetics—while a university student. Then someone told her about Singer's book *Animal Liberation* (1975). After reading the book that converted many thousands of individuals to the cause of animal rights/vegetarianism, she realised something:

I understood that it was much broader than just a cosmetics issue, and that was when I just realised that this was so wrong. I naively then thought all I needed to do was to tell everybody about this and they would all change their lives, as I virtually did overnight, but it didn't happen that way. (Interview 1997)

Nearly twenty years later, she remains an optimist despite the slow progress that the movement has made. She believes in the psychology of small wins, such as a recently won ban on the tethering of pigs in New South Wales and a magistrate's ruling that vindicated attempts by animal liberationists to rescue inhumanely housed battery hens in Canberra. Oogjes believes in focussing on winnable issues, such as bans on animals used in circuses, entertainment, or jumps racing, bans on the face branding of cattle, on the steel-jawed leghold trap, and on some of the more inhumane methods for controlling wildlife. As a self-declared idealist, she is, however, reluctant to put more difficult issues in "the too hard basket," such as live sheep export:

Because we're people that worry about these things, I have great difficulty leaving other issues even if they're not winnable in the short term. If you leave them, they'll never be won. And so we have this background activity on a lot of issues all of the time, for instance live sheep export. We are unlikely to win that for a long time, but people are concerned about it so we try and keep it in front of them. (Interview 1996)

Oogjes cites a recent example where Animals Australia challenged the egg industry, which she believed was using misleading labelling on their products. She wrote to the Australian Consumers Commission and succeeded in preventing the industry from using misleading "happy hen" style images on their cartons:

I love winning stuff like that. And I really think that's a chink in their armour, that they realise that they can't just do anything they want, that there's some public accountability. So I like that sort of thing, and I like being able to empower our people with information. . . . It actually gives me great pleasure to put it into a format, maybe it's a fact sheet, but maybe it's a media release that people can grab on to and can understand. I mean, I really like feeding that information to people and hoping that empowers them. I mean what I get out of the job, is trying to be that conduit. (Interview 1996)

This is not to say that there are no frustrations in her dealings with government bureaucracy. ANZFAS has been providing evidence to most of the Senate Select Committee inquiries and continues to cooperate with government attempts to regulate animal welfare in Australia, although there are signs that the organisation is becoming increasingly disillusioned with the legislative route to animal protection. In a recent paper, the director argued that the Codes of Practice have been used as a public relations smoke screen to soften criticism of the way that animals are treated in various contexts. Oogjes points out that animals of similar or identical sentience in Australia are treated differentially depending on their commercial value. She suggests that to keep a dog in the conditions that intensively farmed pigs endure, as permitted in the codes would constitute an offence and would lead to prosecution. No doubt the apparent inconsistency in the codes is a reflection of the greater emotional value that people accord to companion animals as compared to farm animals.

I asked Oogjes to describe what it was like working as an organisational advocate in the suites and to comment on the intellectual, practical, and emotional dimensions of the work. She explained that she did not consciously think about what the work involved:

> *I just do it, and virtually like anyone else who goes to work, I dislike bits. I work out what things I have to do today, and they are to do with letter writing . . . putting together information, sourcing information, lobbying government representatives, or whatever. I don't sit back and think about it. I mean, there are times where we plan what we might do this year or this month or whatever. And the emotional side of it, I don't even give much time to. That happens when I have to go out to places like abattoirs or whatever. But I see my work more on an academic level and an activist level.* (Interview 1996)

The Ethics of Animal Ethics Committees

An example of the academic work that ANZFAS does can be seen in its submission of more than 130 pages on animal experimentation to the Senate Select Committee in 1987. ANZFAS hoped to improve the treatment and care of laboratory animals. There is nonetheless a dilemma for animal protection SMOs that serve on committees or government inquiries, which are perceived as not able to achieve tangible improvements for animals. Critics maintain that cooperating with the opposition in government and industry legitimates animal exploitation by perpetuating the myth that all is well in the human-animal contract. One critic, who prefers the term "vivisection" to "animal experimentation," asserted that "you can't have a code of vivisection any more than you can have a code for rape" (Coleman 1993:17). As president of ANZFAS, Singer's response is to take a historical and cautiously optimistic view:

> We're still a long way from the goal of ending the institution of animal experimentation and changing the assumption that animals are just tools for us to use. Nevertheless, the codes of practice and the committees do mean that undoubtedly there are many animal experiments which would have been performed in 1975, and which would have caused extreme suffering to animals, which will not be approved in Australia now. This is because of the existence of the ethics committees and because of the animal welfare people on those committees. (1995:9)

The initiative for a code of practice for the care and use of animals in science and research came from the scientific fraternity itself. According to Rogers and Tremont (1992:37), the National Health and Medical Research Council (NHMRC) drew up a code in 1969 as a proactive response to animal rights activism in the 1960s and 1970s. The code is monitored by Australian Experimentation and Ethics Committees (AEECs) and according to Rogers and Tremont, Australia has been spared the radical protests against vivisection and animal research experienced in Britain (1992:37). Yet for animal protectionists, vivisection remains one of the most despised practices involving animals. On a very high to low scale, Animals Australia members rated the use of animals in science and research very high in terms of being morally objectionable. The ranking scores of 1 (extremely wrong) and 7 (not at all wrong) produced the following mean ratings for practices associated with vivisection: Cosmetic/beauty product experiments (1.05); using unclaimed dogs in experiments (1.19); purpose breeding animals for research (1.28); and exposing animals to disease in a medical experiment (1.33).

Clearly, these are practices that Animal Australia members strongly oppose. It is therefore difficult for the organisation to satisfy its membership that its policy of working with the authorities and on ethics committees is anything but an implicit acceptance of such practices. And in fact, some members have left Animals Australia to start their own more activist SMO, most notably in the case of the BUAV and the HSUS. I have witnessed the sometimes very heated debate between ANZFAS members and the leadership on the issue of whether a reformist or abolitionist approach to vivisection is the correct policy for the SMO to take. Because so many of its ordinary members see vivisection as one of the worst forms of animal abuse, many expect ANZFAS to take an abolitionist line on the issue.

At the moment, the organisation's position is to encourage its members to serve on animal ethics committees because they bring a humane perspective to the committees that sometimes produces concrete improvements for animals. For example, some ANZFAS members have suggested that a practise that was permitted five years ago in Australia—the dropping of heavy weights on a sheep's back as part of research on paraplegia—would be rejected today by animal welfarists serving on the committees. In the absence of Australian research to evaluate this claim, we can only turn to the results of two informal surveys by Orlans (1993) of the American equivalent, Institutional Animal Care and Use Committees (IACUCs). The results appear to be acceptable only to the most moderate animal welfarists. Orlans found that it was very rare for any animal experiment to be disallowed completely and was struck by the fact that the only IACUC members to vote against a research protocol were community representatives who tend to have worked in humane societies and the like (1993: 111).

For many animal protectionists, such as the leadership in Animal Aid, the fact most experiments are approved by these committees despite an animal welfare presence is seen as further evidence that an abolitionist approach is preferable to pragmatic animal welfarism. The moderate, reformist approach that is favoured by ANZFAS/Animals Australia is anathema to the next case study in SMO advocacy.

THE UNITED KINGDOM: ANIMAL AID

Animal Aid describes itself as a pressure group established in 1977 to campaign on behalf of farm and laboratory animals. The term "pressure group" is used to distinguish it from animal charities, which are prohibited by the law from political campaigning. Animal Aid, on the other hand, runs peaceful, though confrontational campaigns against cruelty, and it investigates animal suffering and exploitation wherever it needs to in Britain. A cursory inspection of their bimonthly journal, *Outrage*, reveals a broad agenda covering the conventional concerns of animal rights groups, from antifur campaigns to exposing animal suffering in zoos.

Animal Aid's promotional literature points out that it is one of the first national animal rights organisations in the world, and still the only one in the United Kingdom, with a multi-issue agenda. In a supplement to its hundredth publication of *Outrage* (*Animal Aid* February 1996), some of the SMO's highlights and achievements over the past two decades are listed:

1977: Jean Pink, a small-town primary school teacher establishes Animal Aid.

1979: Junior Animal Aid is founded and the first national marches against experimention on animals at Oxford and Cambridge takes place in 1986. The organisation introduces a mock award—the Animal Aid Award for Trivial Research.

1987: Animal Aid's biggest event yet is launched—the Living Without Cruelty exhibition—which 6,000 people attend the following year.

1991: Animal Aid's Humane Research Donor Card is introduced; five years later they claim to have 350,000 card-carrying donors in the UK.

1993: MarketWatch groups are set up to monitor animal suffering at markets throughout the country. (*Animal Aid* 1996)

These few examples indicate the grassroots orientation of Animal Aid, which is based on about one hundred contact groups throughout the country; almost three-quarters of these contact persons are women (Garner 1993:67). The headquarters of the SMO are in Tonbridge, south of London. The group employs about eight full-time and six part-time staff. A recent staff list indicates that there are nine women and six men as well as a council consisting of six women and two men (*Outrage* October-November 1996). The organisation is nonbureaucratic in that it favours a "bottom-up" rather than top-down hierarchial structure. Local campaigners throughout the United Kingdom take the initiative in many campaigns, while the umbrella organisation offers a strategic lead, producing for example, "astute packages on the evils of livestock markets, the sheep and poultry trade, the zoo business, the use of animals in warfare experiments, the squandering of animal lives in tissue testing" (*Animal Aid* February 1996).

Andrew Tyler: Vegan-Animal Rights Activist

As the director of Animal Aid since 1995 and as a professional journalist, Andrew Tyler places great emphasis on these information packages:

> *We have to make a sustained argument. We have to put the facts before people. We have to have access to the media, which is why we have to create eye-catching demos, protests, put out better packages of information and film that the media will be compelled to use.* (Interview 1996)

Tyler's idea of packages of information is much broader than the conventional printed brochures that social movement activists and advocates produce. Animal Aid, like the Animal Welfare Institute in the United States, produces well-researched books and articles on a whole range of issues. Tyler points out that Animal Aid has to target the press effectively with *"intelligent, verifiable packages of information and argument that they can feel secure about using"* (Interview 1996). An example of this is *Silence of the Lambs: An Animal Aid Investigation*, a short monograph that Tyler wrote 1995 that seeks to expose the cruelty involved in the sheep industry in the United Kingdom. As important as these packages are, Tyler believes that Animal Aid must go further to win over the public to its cause:

> *We have to shock and mesmerise and entice, tell powerful stories about the suffering of animals, and what animals really are when they're not molested and confined.* (Interview 1996)

He cites one such story that a sheep farmer's wife tells:

> I was standing on the block the other day and some sheep were coming through and one came running up to me and licked my hands and I said to my husband, "Why is he doing that?" and he said, "You should know, you fed him on the bottle three years ago." They're quite wonderful really, they've got tremendous memories. I can't bear to see a sheep suffering. They don't make a fuss at all, they're a gentle sort of animal, very underrated. And I wonder if it's all worth it really. I ask myself, have we got the right? That's my problem. (Tyler 1995:15)

Stories of this kind are meant to remind people that animals like sheep are individuals with life spans, personalities, and names such as Midnight (aged ten), Fergie (twenty), and Helga (also twenty). Animal rights advocates everywhere use anthropomorphism as a deliberate device to widen people's affection for their pets to include animals lower on the hierarchy of human concern, such as sheep, chickens, and lobsters. Animal Liberation Victoria used the dramatic headline "330 million adolescents murdered in the breeding sheds" to publicise the plight of broiler chickens, whose natural life is about "raking the soil for treasures, bathing rapturously in the dust . . . There's mating games and proud roosters holding court over their flock, nests to build, mothers-to-be religiously warming eggs for weeks, then courageously guarding new born chicks against lurking dangers" (*Action* 1996:8). A similar story is told by Lobster Liberation in the United Kingdom, which asks diners to consider that the lobsters boiled alive for their benefit "have a very long childhood and an awkward adolescence . . . They flirt, their pregnancies last nine months and they can live to be over one hundred 100 years old." What critics would ridicule as sentimental anthropomorphism, animal rights advocates call empathy.

Tyler is adament that Animal Aid remains a nonviolent organisation since he believes that the animal rights movement is motivated primarily by a belief in nonviolence and compassion, adding for good measure the suggestion that

> *Prisons are not filled with rapists and child molesters who are vegan. They are filled with rapists and child molesters who are carnivores. I'm not saying there is a causal link between one or the other, but this is an overwhelmingly peaceful and compassionate movement, and I'm proud to be part of it.* (Interview 1996)

Garner cites evidence of Animal Aid's abhorence of violent forms of direct action and the organisation's condemnation of the ALF's extremism (1993:211–2). Even so, extreme views, if not actions, are suggested in Tyler's observation above. Similar views cropped up in this study, particularly in relation to vivisection and carnivorism. As others have noted in the literature, meat eating has been blamed for a host of social ills, from rape to ecological destruction (Fiddes 1991; Adams 1990).

Animal Aid: Uninhibited Advocacy for All Animals Great and Small

Like most animal rights organisations, Animal Aid promotes vegetarianism and sees meat eating as one of the worst forms of violence against animals and the environment. According to Tyler, Animal Aid is a radical vegan/animal rights organisation that expects its full-time staff to be vegan. He believes that the animal rights movement is unique in its message about living without cruelty, and he suggests that Animal Aid's appeal as a SMO is that it is uncompromising on the rights of animals to be left alone to enjoy their natural lives as sentient (and sensual) creatures, which the above stories illustrate. Tyler explains Animal Aid's appeal as he sees it:

> *We're uninhibited . . . We're not inhibited about saying we want fishing to stop, angling to stop, we want everyone to stop eating meat. We don't put it like that because there's no point in challenging people, doing a guilt thing.* (Interview 1996)

Tyler believes that throughout its history, the animal movement has been characterised by a "them versus us" mentality that has resulted in the movement excluding rather than attracting adherents. And logically, Animal Aid's identity as a radical vegan/animal rights SMO is exclusive, not inclusive, a point which Tyler appears to acknowledge, although he insists on the organisation remaining true to its principles:

But it is important to recognise that we won't get to heaven in a week. We should stick to our undiluted objective, which as far as we're concerned is the abolition of all vivisection, the promotion of veganism, the whole bit. (Interview 1996)

Animal Aid's strict animal rights/vegan identity has, however, not prevented it from working with other more moderate SMOs in the movement, like CIWF, against live animal exports.

And in 1984, Animal Aid and a number of leading antivivisection organisations formed the Mobilisation for Laboratory Animals to campaign jointly against the government's White Paper on laboratory animals. But the animal rights organisation is critical of the tamest and largest of its potential allies in the animal protection fraternity, the RSPCA, which it believes wants only to *"give bigger cages to the captives or lengthen the chain on the slaves."* Tyler points out that his organisation is trying to abolish cruelty and to change fundamental habits. They are not content with the moderate animal welfare stance, which effectively accepts the status quo.

Thus, while Animal Aid will work with any animal movement SMO, Tyler is not enthusiastic about the state of animal protection advocacy in the United Kingdom. He suggests that until the recent live export protests, the animal movement had become stale compared to the late 1980s:

[It was] *young, provocative, imaginative and challenging . . . and did strange things and shocked. And I'm not talking about smashing stuff I'm talking about creative, principled, dynamic action.* (Interview 1996)

Tyler believes that the public and the media are capricious and were stampeded in the direction of live exports in early 1995. But he acknowledges that live exports is an important issue in that it's indicative of the whole spectrum of cruelty to animals, from the forced impregnation to the confinement on the farm to the transportation to foreign places. He also believes that ordinary God-fearing carnivores turned into vegans as a consequence of what they saw during the protests, a claim supported at least in part by Benton and Redfearn (1996: 57). What troubles Tyler is that the export of live animals is only a symptom of the more insidious problem of exploitation of animals. To combat this means challenging the entire system of domination and endemic injustice upon which it is built. This remains the movement's most daunting task, and it is beyond the capacity of any single social movement. Even so, the existence and persistence of Animal Aid and SMOs like it do threaten the status quo:

I have to say our message of radical veganism implicitly does mean radical change to the economic and social structure . . . we are looking for fundamental change, and that appeals in particular to young people who don't have a stake in the system. There's that kind of excitement, that kind of challenge, that's important we recover. . . . And I'm determined that in Animal Aid we recover that dynamic exuberance, and not be inhibited and fearful of losing the support of the small "c" conservative people who think if you step out of line, you want to bring down the state. (Interview 1996)

Hot cognition

Politicians are most responsive to social issues that are likely to incite public outrage. Gamson calls these "hot cognition" issues, since they are laden with emotion (1992a:7). One such issue that highlights this is Animal Aid's dispute with the British Labour Party on its attitude about the use of animals in weapons research. The SMO's 20-city *No Defence* campaign is designed to achieve a ban on warfare experiments, in which 20,000 animals each year are gassed, poisoned, mutilated, blown apart, or subjected to other torments (Tyler 1996:12). Tyler claims that the purpose of the experiments is to mislead the British public into believing that these torments are necessary to protect humans from the devastation of modern weaponry. Tyler says that alternatives—such as studying existing victims of weapons with their consent, human cell tissue, computer simulations, etcetera are already available. Animal Aid used "unorthodox, informal, and above all, critical and value-infused" tactics favoured by social movement activists to attract the media's attention (Crook et al. 1992:155). The *No Defence* campaign included street theatre, an all-night vigil by children and young people outside the Ministry of Defence, a petition that thousands of passers-by signed, posters, banners, and a protest postcard addressed to the leader of the Labour Party. Most people were apparently unaware of the military's use of animals in weaponry tests and were outraged by the idea.

As director of a national organisation with over seventeen thousand members, Tyler is a strong believer in grassroots politics. He is wary of building national groups with large bureaucracies, which he believes can become proprietorial about issues:

That is something the grassroots people will not tolerate. And it's something that we have to work at. And I have to say, for instance, in our bimonthly journal, we do give credit to other groups and their campaigns. We list other groups and what they're doing and encourage people to support some of their campaigns. But there is a certain amount of jealousy and backbiting in the movement. (Interview 1996)

Animal Aid therefore nurtures and listens to its grassroots supporters, and like other successful SMOs, it is able to mobilise its supporters as a pressure group that politicians find difficult to ignore. Furthermore, its status as a pressure group, rather than as a registered charity, means that it can be more "uninhibited" than the RSPCA or CIWF can. Tyler himself is uninhibited about expressing his views on politics in the United Kingdom. While he acknowledges that there are honorable members of Parliament on both sides of the House, he is scathing in his comments about the state of political morality and what he describes as *"a deeply uncivilised culture,"* which he claims, we mistakenly believe is rational and inherently peaceful and social. He is therefore ambivalent about the usefulness of political lobbying *"because political leadership tends to be untrustworthy and pretty hopeless."* The exceptions are the political radicals and dissidents in their own parties. Even so, he acknowledges the value of getting animal welfare issues into the parliamentary spotlight:

> *And yet by doing the political thing, you do attract public attention through the media, and this is in turn builds more public pressure to force governments to act, because governments don't act unless they have to. They don't act because it's a moral thing to do, because it's a principled thing to do. They act when it becomes uncomfortable not to. This is the nature of our democracy.* (Interview 1996)

The final case study of a national multi-issue SMO is the oldest of the three and one that claims to have won over the middle ground to its cause.

THE UNITED STATES: THE ANIMAL WELFARE INSTITUTE

The Animal Welfare Institute (AWI) in Washington D.C. was founded in 1952, and it is one of the oldest national animal protection SMOs in the United States. Like ANZFAS and Animal Aid, the AWI is a small, national organisation with a multi-issue agenda and a high profile in the animal movement. In 1955, AWI set up a sister organisation, the Society for Animal Protection Legislation (SAPL) to encourage people to lobby Congress and state legislatures on animal welfare issues, especially on opposition to the steel-jaw trap, the preservation of endangered species, the prevention of trade in captured exotic birds, the humane treatment of animals in laboratories and factory farms, and the encouragement of humane science teaching in schools.

Christine Stevens, now in her 70s, is still president of the AWI, and she runs the office from her home in fashionable Georgetown in Washington D.C. The first object of interest I noticed was the large Schweizer medal that hangs on the sitting-room wall, which she explained is a replica of the gold original that the institute sent to Nobel Peace Prize winner Dr. Albert Schweizer in 1951. The AWI and its sister organisation now have about a dozen full-time staff members, with slightly more females than males. The *AWI Quarterly* lists an additional seven directors and about 20 officers and members of scientific and international committees.

Christine Stevens: Animal Protection Pioneer

On this last objective, the AWI and its founder, Christine Stevens, are credited with the first important success in America in the reform of science fair standards. Their protests resulted in Westinghouse instituting a ban on the use of vertebrates in experiments at their exhibitions, a ban that has been in place since 1969 (Finsen & Finsen 1994:132). Other more recent successes that the organisation cites include putting pressure on the International Whaling Commission (IWC) to create the Antarctic Whale Sanctuary that protects whales within the sanctuary and getting the IWC to crack down on the smuggling of whale meat. According to Orlans, the AWI and SAPL "have been among the most politically successful of all animal welfare organisations" on a variety of animal protection issues (1993:45). The AWI claims that of the existing laws protecting animals in the United States, fifteen were passed since the founding of SAPL, and "to a substantial extent, because of the Society's work" (SAPL document).

It is instructive to look at some of the federal laws on animal protection since SAPL was founded in 1955. Remarkably, at that time there were only two federal laws to protect animals in the United States. The following selected examples reveal the kind of moderate, incremental advances and hard-fought achievements in the suites that usually do not satisfy the radicals in the movement:

1958: Anesthetisation or instant stunning prior to slaughter except for kosher slaughter.

1966: Minimum standards of care and housing for laboratory animals.

1975: Cyanide collar for sheep, to kill only predators actually attacking sheep.

1978: Humane slaughter of all animals imported to the United States for meat.

1990: Pounds to hold animals for five days before release to dealers.

1991: Exercise for laboratory dogs.

1992: Moratorium on tuna fishing, which kills dolphins.

Purists in the movement would see many of the above as failures rather than as reforms to be celebrated. Concepts such as kosher slaughter, humane slaughter, a cyanide collar to kill other animals, or dolphin-safe tuna are anathema for many hard-liners in the movement. They tend to see animal welfare improvements like a five-day waiting period for pound dogs or exercise for dogs destined for experimentation as nothing more than *"lengthening the chains on the slaves,"* to use Tyler's phrase. What one SMO sees as animal protective measures legitimated and secured in law, another perceives as a sellout.

As founder of AWI, Stevens is acknowledged as one of the most influential forces in changing national animal welfare laws. On the question of vivisection, the reason for which the AWI was established, the Animal Welfare Institute, as its name indicates, is welfarist rather than abolitionist. Stevens believes in working within the system and has done so for almost half a century. The daughter of a physiology professor, she first became interested in animal welfare in the 1940s while working as a volunteer for a pet shelter, where she witnessed the abuse of animals by the proprietor:

My father was the head of the physiology department at the University of Michigan Medical School, so he alerted me to all these attempts by the National Society for Medical Research (NSMR) to get the dogs and cats. Well, naturally, we weren't about to let them do that. (Interview 1996)

Her efforts to stop the shelter's practice of selling animals to research laboratories put her and her sympathetic father into conflict with the National Society for Medical Research (NSMR). Father and daughter offered the American Humane Association (AHA) $10,000 to help them fight the laws requiring shelters to hand over animals to researchers, but they were turned down. The debate that ensued led to dissidents breaking away from the AHA to form the Humane Society of the United States (HSUS) in 1954 (Blum 1994:11; Orlans 1993:45). The controversy also induced Stevens to found her own organisation, the Animal Welfare Institute, in 1952 to challenge the pound laws more effectively. Within a year, an official form the vivisectionist fraternity wrote to her, suggesting that the AWI not join the "group of foolish, misguided, misanthropic, idle, mischievous or mercenary social perverts who form the so-called anti-vivisection societies to a large extent" (Blum 1994:112). Apart from making her angry, these sentiments were a warning to the AWI that it was in for a long struggle with medical researchers over their (mis)use of laboratory animals. In 1960, after preventing researchers from seizing pound animals for their experiments, the AWI campaigned for the inclusion of laboratory animals in the law to protect animals from abuse.

On her father's advice, Stevens visited the United Kingdom, where animals had been protected since 1876 by the British Cruelty to Animals Act. She recalls meeting scientists who said they were happy to work under the act since it prevented unnecessary cruelty and encouraged the right attitude towards animal experiments. Among the distinguished English physiologists she met was Sir Alexander Fleming, who showed her the windowsill on which he had discovered penicillin. She still remembers Fleming's rabbits in spacious cages, with room to hop about and sit up. After her return to the United States, she continued to fight for legislation similar to the British anticruelty act.

It was not until 1966 that the United States received its first comprehensive animal welfare act, and according to Blum this was due to the singular efforts of Stevens, whom she describes as "both a strong and a rational voice in the animal protection movement" (1994:110–111). Stevens herself is more modest. In a speech to the United States Department of Agriculture (USDA) to mark the thirtieth anniversary of the Animal Welfare Act of 1966, she told the participants that success was due to the excesses of the random source dealers themselves, who continue to be the source of much animal suffering (1996:14).

On the continuum of animal protection SMOs, the AWI is situated somewhere in the centre. Garner (1993:52) describes the organisation as having "a typical welfarist outlook," a description that Stevens confirms in her outline of the genesis of the organisation:

We were founded because of the fight between the medical extremists and the antivivisectionists who say, "No experiments at all." And then the NSMR [National Society for Medical Research] *said, "No holds barred on any kind of experimentation," so nobody was doing anything for the animals. The animals were left out of this. They just enjoyed their philosophical contest, and of course, the antivivisectionists were not accomplishing anything, alas, at all. And so we came in and said, "All right, we're the middle ground, and we believe that it has to be controlled and regulated—reduced numbers, decent housing, decent treatment, all of those things.* (Interview 1996)

As noted earlier, these improvements in animal welfare are not celebrated with the same enthusiasm by more ardent animal rightists in the movement. Nonetheless, Stevens knows better than most what is achievable and what is not:

We feel that, usually, when there are polls taken, that our policies are about the same as the opinion of the average American . . . And the results, when we succeed, are what the American public wants. Their general outlook is quite similar in most cases to the Animal Welfare Institute. (Interview 1996)

Intellectual Work at the Animal Wefare Institute

The Society for Animal Protection Legislation (SAPL) has a long list of achievements since it was set up in 1955, not the least of which is the documentation that contributed to the Animal Welfare Act of 1966 and the widening of its provisions in 1970, 1976, 1985, and 1990. AWI and SAPL are unique among animal protection SMOs in that much of their funding comes from the production of high-quality publications, which also serve as resources in their legal work. Some of the publications are provided free to other animal welfare bodies as well as to public libraries, police departments, and scientific institutions. While most of the big, well-funded animal SMOs produce impressive monographs and glossy magazines, none rival the AWI in terms of the range and depth of output. Considering the relatively small number of staff employed at AWI, this is an impressive achievement. The list of publications and videos includes twenty substantial books and monographs averaging 165 pages in length on issues like animal appreciation and protection, wildlife, trapping, laboratory animals, factory farming, and humane education. In its campaigns on these issues, the AWI uses knowledge as an organisational weapon in much the same way that ANZFAS does in its lobbying and advocacy work.

Campaigning Against Leghold Traps

For the past several decades, the AWI has documented the cruelty involved in the trapping of wild animals. *Facts about Furs* (Nilsson 1979), by AWI staff, has been an influential weapon against the fur trade since it was first published in the late 1970s (Jasper & Nelkin 1992:150). SAPL also used the actor Gregory Peck in its video exposé of the fur trade, *Gregory Peck Speaks Against the*

Leghold Trap, a device against which AWI is still vigorously campaigning. Much of its intellectual work has gone into this issue in the past few years as a result of threats to the European Union's ban on the leghold trap in 1991.

The European Union also placed a ban on the importation of furs from countries that are not among the more than eighty nations that have banned leghold traps. Australia and New Zealand still use steel-jawed leghold traps to control "noxious" animals as well as for hunting and research purposes. Canada and the United States oppose the ban and are seeking to challenge the European Union's regulation through the World Trade Organisation arguing that the ban breaches the free trade agreement. Writing in the AWI *Quarterly* (Stevens 1996b), Stevens called for the United States "to join the rest of the civilized world. In the name of human decency, our country must stop defending the indefensible . . . Every red-blooded American should revolt against our government's hypocritical defence of the steel jaw leghold trap." As indicated in Table 8.1, American and Australian animal protectionists ranked the device as the worst abuse in a list of 15 practises involving animals.

In the best tradition of animal welfare advocacy, AWI's strategy has been to challenge the supporters of the trap with an education campaign designed to expose the cruelty of the device and to suggest alternatives. It commissioned a wildlife expert to create a series of technical monographs on alternatives to the leghold trap in 1990 and has made updates available to interested parties. Influential groups were lobbied in the United States, Canada, and Europe, where AWI's executive director, Cathy Liss, visited the European Parliament and the European Commission to emphasise the strong public support in America (78 per cent according to a Yale University survey and 74 per cent according to the AWI's own survey) for Banning the leghold trap. She also informed the policymakers of the more humane alternatives that are currently available. Banning the leghold trap is close to Liss's heart, and she has written about the issue in *Animals and Their Legal Rights* (1990). She has worked at the AWI for fifteen years, and banning the leghold trap has been one of her main preoccupations. In the following extract, she reflects on the prospects for getting the trap banned in the future:

> *I guess trapping is probably one of the first issues that I worked on when I came here. And like most people, I said, "They're still using that out here?" It was just beyond us because it's not something that I think people encounter in their day-to-day lives. Most people are horrified that it's still used, and were they to see the result, I think would be appalled by it. And I thought, "Of course, we'll be able to get rid of this in no time." And naively, I just couldn't imagine otherwise. And here I am fifteen years later, still saying, "My God, am I going to see this banned in my life time?*" (Interview 1996)

Hard-line animal rights proponents in the movement would undoubtedly argue that Liss's disappointment is the result of a moderate, reformist approach that inevitably produces only incremental improvements in animal welfare, or none at all. According to Jasper and Nelkin (1992:85), only a few states in the United States have responded to peaceful lobbying; in 1983, a bill that sought to ban fur from animals caught in the steel leghold trap was defeated despite having 125 sponsors. The animal movement is divided, however, on the kind of tactics needed to seriously stigmatise traps and furs.

But the real challenge confronting the AWI, according to Stevens, is not the campaign against cruelty, but economic factors. She quotes the opinion of the venerable RSPCA (UK) on the European Union's regulation against the leghold trap: "It appears that trade considerations have meant that a piece of legislation agreed on over four years ago has been superseded by the mere hint of a challenge to free trade. This is a pessimistic precedent for the implementation of future animal welfare legislation" (Stevens 1996b:4).

Whether the issue is the leghold trap, vivisection, or factory farming, animal advocates attempt to make different forms of knowledge count. Each of the SMOs in this chapter is unique in the way that it practises animal protection. As animal protection organizations with different ideological stances—animal welfare (AWI), animal liberation (ANZFAS), and animal rights (Animal Aid)—they share the same philosophy of compassion and justice for animals that is the linchpin of the animal movement worldwide. However, each SMO has its own identity and uses different strategies and tactics to achieve the movement's common goals.

ANZFAS's pragmatism is closer to AWI's middle-of-the-road approach than to Animal Aid's more confrontational style of animal rights/veganism. These orientations have much to do with the leaders of the respective SMOs. Singer's philosophy of utilitarianism enables ANZFAS to operate pragmatically as a regional federation made up of member societies whose identities encompass the entire range of animal protection. This would not be possible nor desirable in an SMO like Animal Aid, whose abolitionist position calls for a moral entrepreneur of Tyler's persuasion. His strict radical vegan/animal rights philosophy clearly identifies the SMO as a distinctive pressure group supported by its grassroots and respected by like-minded activist elsewhere. The Washington D.C.-based Animal Welfare Institute, with its political lobbying arm SAPL, is very much a creature of its founder, Stevens, who shares Tyler's passion and Singer's pragmatism, although ironically, not Singer's faith in philosophy. She is dismissive of philosophical debate and sees more value in securing concrete reforms and improvements in animal protective legislation than in endless discussions about "rights," "ethics," and other philosophical abstractions. Even so, the AWI, more than most other national multi-issue SMOs in this study, uses knowledge as an organisational tool in its animal advocacy work.

Tactically, the three SMOs are similar in the reportoire of actions that they adopt. Although Animal Aid is more unorthodox than ANZFAS or the Animal Welfare Institute is, it espouses and practises nonviolence in its campaigns and advocacy on behalf of animals. And finally, whatever the differences among these three exemplary SMOs, they each use knowledge as an organisational weapon to achieve improvements in the way that we treat animals, especially in the contexts listed in Table 8.1 in the early part of this chapter. Most of these issues will be discussed in the remainder of the book. In the next chapter, "the difficult issue" of animal experimentation is discussed.

NOTE

1. Questions used in the ASIS instrument that were borrowed from Richards' (1990, 1992) survey in the United States offer some interesting comparative data on attitudes toward animals. Respondents were asked to rate on a point scale, "extremely wrong (1) to not at all wrong (7)," the fifteen human uses of animals. The mean scores for both samples indicate an identical ranking order by respondents in Australia and in the United States, as shown in the table.

Chapter 9

Scientific Antivivisection and the Liberation of Laboratory Animals

It is the rarest thing in the world to hear a rational discussion on vivisection.
—C S Lewis

In the above remark, which introduced his pamphlet *Vivisection* in 1947, C S Lewis expressed what many others have continued to assert some fifty years later, namely that the adversaries on both sides of the animal experimentation debate appeal principally to emotion in making their respective cases. Lewis argues that a rational discussion of animal experimentation must begin by asking whether pain is, or is not, an evil: "If it is not, then the case against vivisection falls. But then so does the case for vivisection" (1947). This is the heart of the ethical dilemma concerning animal experimentation that continues to fuel the acrimonious debate between animal researchers and their critics.

Fifty years after Lewis wrote his piece, a cover story in *Scientific American* in February 1997 canvassed the views of scientists on both sides of the debate. Andrew Rowan introduced the forum by noting that the past twenty years have witnessed a "largely unproductive debate" about the pros and cons of animal research (1997:79). Elsewhere I have argued that the issue is still unresolved because of the futile search for the middle ground between animal experimentalists and animal liberationists (Munro 1999a). Whenever an issue is framed as "two-sided"—as in the recent book *Brute Science* by LaFollette and Shanks (1996)—there is the temptation to find a middle way. As the history of opposition to animal experimentation has shown, this has not been achieved since the debate on both sides has been characterised by vilification, rather than by accommodation.

The term "evil" is part of this rhetoric of vilification. As I explained in Chapter 1, from the very outset, opponents of animal experimentation have denounced vivisection as an evil, not least because civilized professionals do unspeakable things to sentient beings.[1] The same tone of condemnation appears in antivivisectionist declarations in the late twentieth century, as the following excerpt shows:

> Behind laboratory walls all over the world, Science (men and women paid for their scientific work) drills, incises, chips, injects, inserts, cauterises, lesions, sutures, amputates, paralyses, tests, deprives, rewards, jabs, and shocks legions of living creatures which are more often than not fully conscious, ie. not under anaesthesia and post-operative sedation. (Collard 1988:60)

A half century after the publication of Lewis's pamphlet, it is noteworthy that the word "pain" did not rate a mention in the case for vivisection that Botting and Morrison, the two animal researchers, made in *Scientific American*. They instead focussed on the necessity and relevance of animal experimentation for human health. The only acknowledgement that animals, as sentient beings, might suffer pain and distress in the process was implied in their blunt claim that "there are no basic differences between the physiology of laboratory animals and humans" (Botting & Morrison 1997:85). If this is the case, then animals, like humans, suffer pain when they are subjected to the kinds of procedures mentioned in the article. For example, the authors refer to Pasteur's experiments on rabbits and guinea pigs that "obviously . . . could never have been carried out on humans" (1997:83).

In presenting the antivivisection argument in *Scientific American*, Barnard and Kaufman also emphasised pragmatic rather than ethical arguments. When they did raise the issue of animal pain, it was indirectly referred to as a factor that distorts experimental results. Cruelty was mentioned only in the form of mental cruelty when they referred to Harlow's monkey experiments in the 1960s, "which left the animals severely damaged emotionally" without any scientific advance (1997:82). In their condemnation of animal research, the antivivisectionists focus on the inapplicability of animal tests and their inferiority to many of the available alternatives.

In both the arguments for and against vivisection, the pain and distress of laboratory animals—the basis of C S Lewis's position—are seen in this contemporary debate as less salient than are the outcomes for humans, either as beneficiaries or as victims of science. Thus, for both animal researchers and their opponents, the issue of cruelty to animals has largely been superseded by a concern for human well-being. And as I describe below, this frame extension has been replicated in the animal rights movement itself, often producing the same acrimony that characterises relations between vivisectors and their opponents.

According to LaFollette and Shanks, the weight of public opinion is largely on the side of the scientists since animal experimentation continues in countries where the animal liberationists and antivivisectionists are most active. The authors suggest that the research community has been more effective than their critics have in presenting a prima facie case for animal experimentation (1996: 17–18).

What is more likely, however, is that the scientists have speciesism on their side. For when it comes to resolving the problem of pain, people are inclined to put themselves and their species first, despite the fact that public opinion does seem to be against animal experimentation. A national survey in the United Kingdom in 1989 found that half the respondents believed that the practise was morally wrong, which was more than the number claiming to be against abortion, capital punishment, or pornography (Plous 1996:1169). Similar results attesting to public opposition to animal research have been found in several countries, including Australia, according to the Australian Association for Humane Research (AAHR). Yet it has to be said that "public moralism" rests on the very shaky foundation that underlies the norm of *moderate* concern for animals. Self-interest, even for someone strongly committed to the principle of doing no harm to animals, will usually be given priority when more than trivial health considerations are at stake.

THE SINS OF ANIMAL EXPERIMENTATION AND ANTHROPOMORPHISM

According to prominent British philosopher Stephen Clark, animal experimentation "represents our single greatest sin" against "our kindred" (1997:154–168). For Clark, the sin is the intention always to control how things work, to evade all penalties, to never yield to nature. He argues that the experimentalists' sin is not that they are cruel or negligent, but rather that they act out false or wicked assumptions that are deeply ingrained in contemporary culture. Clark believes that moral orthodoxy, or what he calls "public moralism—the norm of *moderate* concern for animals," is morally bankrupt. Moral orthodoxy means that if animals do not suffer unnecessarily, they can be dissected, eaten, and used to satisfy human needs and wants.

As I explained in Chapter 3, this pragmatic, welfarist approach is advocated by R. G. Frey, who proposes the notion of "the concerned individual" as a model for an ethical treatment of other animals. Clark also proposes the concept of "an ideal jury" as a means for achieving a disinterested evaluation of the aforementioned assumptions. Just as Frey's concerned individual would support the norm of *moderate* concern for animals, Clark's disinterested jury members would find it wanting. Clark's verdict is that we have no right to to use any creature in ways that would impair its chance of living out a natural existence. He believes that an ideal jury would not condone animal experimentation and would condemn the rhetoric of modern objectivism as morally impoverished. Furthermore, he suggests that those scientists "who insist on seeing laboratory animals as mere 'animal preparations' or living test-tubes deny themselves the chance to see such truths as would cause difficulties for their treatment of them" (1997: 167).

Lynda Birke has reflected both on her own occupational socialisation as a scientist and on what is involved emotionally in the daily practises of science involving the use of animals. One of the things the neophyte scientist must learn, she claims, is to fit into the "macho culture of the laboratory," which entails avoiding "the sin of anthropomorphism." To express anxiety about the suffering of laboratory animals is unscientific and comes dangerously close to animal rights rhetoric. Birke notes several contradictions in the way that scientists perceive animals. For example, anthropomorphism is evidently acceptable in the case of pets, but not for animals labelled and numbered as "research tools" (Birke 1997).

Feminists like Birke are proud to embrace anthropomorphism as an emotional resource. Sara Ruddick, for example, would like scientists to be trained to empathise with laboratory animals by asking their captives, "what are you going through?" Only by reflecting on the visceral rather than on the objectively intellectual can scientists hope to learn an ethic of care and humility, thereby replacing a scientific epistemology with a maternal one. But this, she suggests, would spell the end of vivisection, and so emotional responses to animals are ridiculed as unscientific. However, this notion of the unfeeling scientist's conversion to a new ethic of respect for animals is far-fetched and, like similar moral dichotomies, overly simplistic. The culture of animal experimentation that colours the world of the vivisectionist is a shade greyer than this black-and-white image suggests. As one of a handful of sociologists studying the place of animals in human society, Arnold Arluke's analysis of the work of animal experimentalists has revealed the complexity of animal workers' relationships with laboratory animals. His studies suggest that these workers are not all similar in the way that they perceive and treat their research animals (1990, 1991).

The conflict between the comfort and physical well-being of human beings and the ruthless exploitation of other forms of animal life, to paraphrase Keith Thomas (1983:303), applies to most human practises involving the lives of animals. Nowhere is this conflict more evident than in the clash between liberationists and experimentalists over animal research.

VIEWS FROM INSIDE THE ANIMAL PROTECTION FRATERNITY

Animal movement observers usually identify animal experimentation and research as the most bitterly contested issue in the movement because of the ideological gulf that separates the protagonists in the conflict (Harris 1985:170).

And for many animal activists and advocates, it is *the* most divisive issue in the animal protection movement. Time and again, what Ritvo called "the hardest case" (1987:157) was referred to as a difficult, intractable issue by interviewees in the three countries in the present study. And for many people inside or outside of the movement, animal experimentation remains the most important moral dilemma, as well as the most controversial question, for the movement to resolve.

Movement activists and analysts agree that animal experimentation is a deeply emotional issue (Groves 1995, 1997). As the data in this chapter make abundantly clear, it is "a hot cognition" issue, to use Gamson's phrase (1992). We will begin with the interviews I conducted in the three case study countries. What is immediately apparent in the excerpts is that the issue seems impossible to resolve to everyone's satisfaction.

Animal Experimentation = Hot Cognition

Sue, an Australian animal liberationist, explained why vivisection was a hot issue:

I think it's an intractable issue, but I was thinking mainly in terms of the numbers of animals involved . . . and we don't have any scientists in this country. The movement is not yet big enough to be able to pay the scientist to desert the lab and come over to our side and do the research without animals. (Interview 1997)

I wondered if there was any area, such as veterinary medicine, that animal rights activists would exempt from moral condemnation. The intractability appears to be on both sides, as disclosed in the response to the question that I put to a staff member at the British Union for the Abolition of Vivisection (BUAV):

L M: It struck me that some vets probably do experiment on animals, but not for the purposes of human well-being, but for animal well-being. Do you think that's fair enough?

Kaye: *No I mean, if you're developing a dog vaccination against a disease then, no, I don't think it's right that one dog should have to suffer for the sake of these other dogs. We don't do that as human beings. We don't have guinea pig-human beings. So no, especially because the animals that they're trying to protect are people's pets, or even a lot of the animal research that's done is for the wrong reasons. It's for profit, it's for developing horses that will race faster or recover from injuries quicker. It's all about money.* (Interview 1996)

The president of Animal Liberation Victoria, Rheya Linden, took a similar abolitionist stance:

Sometimes there is valid information, I don't want to deny that. I mean, whatever you're doing you'll get some valid data. . . . I think fundamental to that, even if it were proven to me beyond any reasonable doubt that it was the most useful form of data collection available, I would still have to say no because we do not have a right. (Interview 1997)

And later in the interview, she adds that

I think it's really important to engage in dialogue and debate because unless we can back up our protests with good arguments, and they're not necessarily animal welfare arguments, they're also scientific arguments, as far as we have the same scientific expertise. (Interview 1997)

How then is it possible for Linden to "*engage in dialogue*" if her purpose is to ban animal experimentation all together? Moral entrepreneurs in social movements seek to change society, not just talk about it. Linden, in the classic style of moral entrepreneurship, "aims to create categories of deviance through legal and attitudinal change" (Roach Anleu 1993:53). She vilifies the vivisection fraternity by association when she alludes in the interview to examples of atrocities that scientists inflicted on humans in South Africa under apartheid and in Nazi Germany.

As president of Animal Liberation Victoria, her activism is primarily directed against vivisectionists, although she is ambivalent about science and scientists in general:

We are advocating for a vulnerable population, whether it's animals or humans. And on the other side of it, we conclude, as I do conclude, that science works for itself, that science does not work with a morality. And despite ethics committees, we know that in the end what matters is the grant and the status . . . It's like a heroic race to get there first and this is what empowers them. (Interview 1996)

Later in the interview, she elaborates on what she sees as the competitive culture of the scientific enterprise:

They [the scientists] are still the elite. I mean, it's often said that science is a kind of class A religion . . . and hearing scientists talk about their research only confirms that for me. Witness the excitement of being there first, this global race to get the cure, the first cure, or the first this and the first that. And I mean, it's really very frontier-building, and like the quests of old I guess. It's very masculinist as well. . . . I think that nothing can stand in the way of that, and we need to unlock that, we need to have it much more open. The taxpayer who funds this race needs to be able to see what's being done [in the name of science]. (Interview 1996)

The American animal protectionist, Tamara Hamilton, who at the time of the interview worked for the HSUS, does not see vivisection as an exclusively "masculinist" vice but like Linden, she sees the competition for science research grants as a deeply flawed process based on a quest for "*the almighty dollar.*" While less judgemental than most antivivisectionists I have encountered, one of the hardest things she said that she had to deal with in her job was the attitude of scientists that

If you're not a research scientist, you couldn't possibly understand. And it's hard for them to see your side of it, and be able to get them to see that they don't see the animal welfare portion of it. They don't understand that. That's very difficult to convey to them, throughout their whole lives they've been conditioned to use animals as research tools, and to try and get them to see them as something other than as a resource is difficult. (Interview 1996)

Hamilton had impressed me with her passion and commitment to the animal welfare cause, and I wondered about the intellectual and emotional aspects of her work. The printed transcript does not convey the strength of feeling in her reply, which I remember as one of the most sincerely held responses in the entire study:

Certainly, emotions are an important part of it, because we have our hearts involved. You need to keep your heart in something, I think, to be effective. And if it's something that you believe in, you'll be more effective. (Interview 1996)

Hamilton explained that the issue of laboratory animals was the one thing that she would like to see resolved for the benefit of animals. She considers herself privileged to be playing a part in the movement:

The use of animals in laboratories, for me, is very near and dear to my heart. I would have come to work for the HSUS in any capacity . . . I know so well what the conditions are in the laboratories . . . Like most scientists, I would like to see the end of research that harms animals. I would like to see the day when there are no animals used in research. (Interview 1996)

In her early thirties, Hamilton does not believe that day will come in her lifetime. She mentions the power of organisations like the NABR and Americans for Medical Progress that keep much of the animal research going and keep the animals in conditions that serve commercial interests rather than the interests of the animals.

Like many animal protectionists in this study, she had a visceral experience that provided the catalyst for her initiation into the animal movement. She explained how she accidentally came across a piece of mail addressed to the National Antivivisection Society. On the front of their literature was a picture of rabbits. As she had been given a stray rabbit a couple of years before, the picture and what it stood for came as a shock:

It had touched something in me that all of a sudden, they weren't doing this just to rabbits, they were doing this to Edison, who was my rabbit. They were doing things to

Edison, and so then I was very concerned because if it would hurt Edison, it would certainly hurt the other rabbits too. (Interview 1996)

While rabbits do not enjoy this kind of affection in Australia, animal protectionists from welfarists to abolitionists object to the experimentalist's labelling of animals—including rabbits—as "research tools" and the like. This is not an emotional issue, as some critics of the movement allege, but rather an issue of social justice. For as long as animals are talked about as "resources" or "property," there is little chance that they will be treated as sentient beings.

One of the Australian activists made this point during the interview, in reference to the ideas of the animal rights advocates Regan and Francione. At the time of the interview, Owen had been working as a volunteer of Animal Liberation NSW for about eighteen months. He is a Sydney University student enrolled in science and specialising in math and computers. I asked him if he had any views on the idea of purpose breeding animals for research instead of using strays or pound dogs, cats, or rabbits for veterinary research. He did not support the idea and stressed that animals should not be treated as property, adding that science has no right to play with human life:

If I purpose bred a child, and if I could somehow genetically engineer him so he was incredibly retarded, it's still wrong. (Interview 1997)

I wondered what the point of such an experiment would be. Months later, I came across Clark's startling observation of "the real possibilities" for scientists "to breed pretty imbeciles for licensed pederasts" (1997:164).

Not all animal protectionists were as suspicious of the scientific fraternity as some of the individuals quoted above. In the ASIS questionnaire, only 25 per cent of the sample agreed with the proposition that scientists do more harm than good, compared to 21 per cent who disagreed. Ambivalence characterised what many animal protectionists who were interviewed or responded to the survey believed about science. One Australian activist/advocate I interviewed, Lisa, expressed this ambivalence when she spoke warmly of a number of scientists she knew who cared about animals, but referred in particular to one researcher's good husbandry and ostensible love of animals that was diminished because

She experiments on them . . . When it comes to the crunch, they can cut off their emotions and distance themselves from it, whereas people in the animal rights movement can't, can't cut our emotions off like this. And we find that we can't distance ourselves from the problems. (Interview 1992)

The role of emotion versus reason is prominent in the animal experimentation debate. I asked Dr. Gus Thornton, CEO of the MSPCA in Boston, if he believed that emotion is something that can be used positively in animal protection work:

Oh, of course, I would say yes, it can be a positive or it can be a negative, depending upon the degree or the amount of control one uses. In other words, in my opinion, if people become totally nonpragmatic, nonrational and are totally controlled by their emotional feelings, I think they're very ineffective, incredibly ineffective long-term. Because I think that if your concern is the animal and its welfare, you have to be willing to work in the real world, where real things happen. Animals are eaten and animals are killed. There are too many companion animals, and many have to be destroyed because of that. (Interview 1996)

This is the classic welfarist perspective that most of the aforementioned animal protectionists would disavow as too conservative. Finally, I put to the head of the MSPCA—which employs about sixty people, more than half of whom are women—the idea that in the future, the veterinary profession will be influenced by an increasing number of women joining the profession. Without wishing to generalise too much or to exclude men who cared about animals, he believed that *"females tend to be more empathetic, sympathetic, more inclined to address animal welfare issues."* . . . and *"relate in some instances better to the distraught owner."* In the next section, I discuss the attitude of women to animal experimentation and the role of caring in the antivivisection movement.

Female Attitudes Toward Animal Research

Women, as we saw in Chapter 1, were prominent in the early protests against animal experimentation. And as I noted in Chapter 4, women greatly outnumber men in the contemporary animal liberation movement. In addition, several studies have shown that women are more ambivalent than men are about the putative advantages of science and technology to human well-being. One recent study concluded that there are significant gender differences in attitudes toward the use of animals in research (Plous 1996). Plous surveyed some four thousand psychologists in the United States and found that female respondents were significantly more sympathetic than were their male counterparts toward animal research, defined as psychological research, not biomedical or toxicology testing. While the study's general conclusion was that animal research in psychology enjoyed qualified support from the respondents, female psychologists differed significantly from their male colleagues on eight of the nine measures.

How can these differences in attitudes between the sexes be explained? According to the ecofeminist Andree Collard, the relevance of animal experimentation to women's lives is unmistakable; she sees male power as the connecting link between the oppression of women and animals. She claims that "the banality of cruelty in patriarchal cultures" is responsible for the continuation of "humane" animal experiments at "safe/moral" levels, just as it is responsible for the abuse and exploitation of women (1988:73). Collard gives the example of caesarian surgery, which technicians and doctors perform on rodents and women. Because male doctors are unable to give birth themselves, they stimulate "the power of female generativity by manipulating her processes with drugs, forcing her to yield to *his* time, and threatening and/or taking her very life" (1988:74). The only difference between the pregnant human and nonhuman is that the rodent is automatically killed/sacrificed. The justifications given for the sadism of animal research, she suggests, are reminiscent of those that rapists offer for male sexual violence against women.

To Collard, science is male-dominated and sinister. In reconstructing sentient creatures as "living-tools," the researcher is provided with a crutch, "whereby he can achieve the patriarchal ideal of godliness: power to destroy, power to control" (1988:66). This view of science cannot be easily dismissed as the ravings of a lesbian man-hater as it was held just as firmly by "respectable" antivivisectionists of the nineteenth century. Within the mainstream animal protection movement, there is some evidence that women are more ambivalent about science and scientists than men are. My own survey (ASIS), for instance, revealed that on a number of issues associated with animal experimentation, women registered significantly higher levels of disapproval than men did. In one question, which suggested that on the whole, scientists do more harm than good, female respondents had a significantly higher mean than males in agreeing with the proposition. These two views of female attitudes toward science—ranging from outright hostility to wary ambivalence—spring from the same source, the idea of women and nature as victims of science's unrelenting quest for mastery over what Schiebinger called "nature's body" (1993).

Given these views of science as antithetical to the interests of women, how can the position of female scientists be explained? Birke points out that science is quintessentially about the pursuit of objectivity and therefore "denies feelings, including the possibility of feelings of sympathy towards the objects of study—be they nonhumans, the environment generally, or women" (1995a:43). Feminist scientists have begun to challenge the "otherness" of the subject matter of science, as Birke herself does in her various writings (1986, 1994, 1995b). But as Birke suggests, the boundaries of "otherness" are now being blurred within mainstream science itself in the form of organ transplants, the transfer of genetic material from one species to another, and in the creation of transgenic mice and rats. As a consequence of these developments, it is increasingly more difficult, she suggests, for scientists to be able to relegate laboratory animals to the realm of "otherness."

One of the most important arenas for women to practice science is in the veterinary field where, unusually, they outnumber men. However, Tannenbaum (1993) points out that he could locate only three studies by social scientists on the moral development of vets in the United States, and these were of limited value. Information of a statistical kind paints a more promising picture for the professional treatment of animals. In Australia, for instance, in the class of 1995–1997, the proportion of female undergraduates in veterinary science was about two-thirds. Tannenbaum has also noted the increasing feminisation of the veterinary profession in North America, where the number of females in some courses exceeds 75 per cent (1993). He argues that this trend, along with the increased attention to formal ethics training evident in the veterinary curriculum, will lead to greater ethical sensitivity in the profession, a point on which Thornton agrees (Interview 1996).

On the broader front, feminist principles, in so far as they are expressed "in a different voice" (Gilligan 1992), can be used to explain the oppression of animals in patriarchal societies. Gilligan points to a difference between feminine and masculine modes of thinking; women tend to be concerned with obligations and responsibilities, and men tend to be concerned with the more abstract concepts of rights and justice. The socialisation of men involves emotional distancing, while for women, emotional attachment is the norm; these different patterns of socialisation encourage different moral orientations between the sexes. Women, more so than men, are likely to favour a policy of inclusion and caring for others, including animals. Even the prominent animal rights/liberation philosophers Regan and Singer have been taken to task by feminist critics for ignoring emotion and sentiment in favour of male reason and rationality. Sara Ruddick (1980), for example, believes that the scientific epistemology practised by these philosophers needs to be supplanted by a maternal epistemology based on an ethic of care and humility. To do this, insists Josephine Donovan, feminists must "reject carnivorism . . . the use of lab animals for testing of beauty and cleaning products (such as the infamous LD50 and Draize tests) and military equipment as well as psychological experimentation . . . and supports efforts to replace medical experiments by computer models and tissue cultlures" (1993: 184).

Thus, feminist principles, so defined, will be as important to the animal movement in the future as they have been in the past, although many feminists do not agree with "the maternal epistemology" concept or the need for a conversion to vegetarianism (George 1994). Furthermore, as we saw from the demise of the early animal protection campaigns in Australia, there are dangers for a movement that relies too heavily on expressive, nurturing orientations at the expense of a political activism that includes the reforming zeal of both men and women.

Successful social movements need more than a handful of committed activists. As we have seen elsewhere in the book, individual moral entrepreneurs are important catalysts for change, but the movements that they inspire are likely to peter out if they depend too heavily on a few charismatic individuals. This is not to denigrate the efforts of the individual activist or the power of one, but rather to suggest that Amnesty International is more effective in the long run, than say, Noam Chomsky. In the next section, the example of the late Henry Spira suggests that it is possible for an individual to mobilise mass support for social change. However, long-lasting change will only be forthcoming if movement organisations are built that outlast the typically brief tenure of the individual movement entrepreneur. Elsewhere (Munro 1999a), I have described

Spira's campaigns, and the advantages of collaborative coalitions between like-minded groups and individuals that Spira favoured, in more detail than is possible here.

Social Activism, Henry Spira-Style

For many people, psychological research involving animals seems more benign than biomedical or toxicology tests, where invasive procedures are used more extensively and routinely. Singer (1990:42) claims, however, that many of the most painful experiments are performed during psychological research, and he is scathing about the futility of many of them. One such experiment became a *cause célèbre* in the modern animal liberation movement in the United States when Spira exposed the dubious science of psychological experiments on cats at the American Museum of Natural History in New York. The procedure involved destroying the natural bodily functions of the cats to study the effects of mutilations, such as brain and penis damage, on the animals' sexual performance. Spira condemned the experiments as cruel and pointless, and he had little difficulty in gaining public support for their termination after it was revealed that they had been going on at great public expense for some seventeen years. Spira's victory over the museum in 1977 represented a turning point for American animal protectionists, who for the first time witnessed the success of a focussed, grassroots, activist campaign. It was also the first time that an activist had invoked animal liberation theory against science. Spira maintained that animals were not things but sentient beings with feelings, and scientists had no right to inflict pain and suffering on what many of them callously labelled "research tools."

Singer has since acknowledged the importance of Spira to the animal movement in his documentary film, *Henry—One Man's Way* (1997), and in his book *Ethics into Action* (1998) on the life of the late activist. While this is not the place to report in detail on the Singer-Spira partnership, it is important to point out that Spira's "fateful moment" in his long career as an animal activist came after he attended a lecture by Singer in New York in 1974. This was also after he had read the speaker's early ideas on animal liberation, published in the *New York Review of Books* in April 1973. This is a classic instance of the powerful alchemy of animal advocacy based on moral philosophy and animal activism, and of the use of public protest via the media to galvanise support for the cause. Spira's museum victory encouraged many other activists to organise themselves into social movement organisations, such as People for the Ethical Treatment of Animals (PETA), which is now one of the world's largest animal rights groups.

Spira used an approach similar to "reintegrative shaming" (Braithwaite 1989) that proved more effective than the stigmatising or disintegrative shaming of the earlier antivivisectionists did. He not only moralised with alleged animal exploiters, but offered to work with them to find more acceptable alternatives to the Draize and LD50 tests. More radical animal rights groups, like PETA, were appalled by Spira's support of companies that had a history of exploiting animals. Spira's response was that his strategy of accommodation, a version of reintegrative shaming that favours reinforcement and forgiveness, leads to less animal suffering and is more effective than the vilification and stigmatisation of opponents is.

On this he is surely correct. The abolitionists' rhetoric is usually counterproductive, as is their negative approach to shaming "animal abusers." Stigmatising the "criminal," especially white-collar offenders, does not appear to be as effective as moralising with them and encouraging them to change their deviant ways. According to how the theory of reintegrative shaming works, the crime, not the offender, is the focus of the moralising effort. Spira appears to have used this technique effectively a decade before the Australian criminologist Braithwaite formulated the theory as a means of controlling crime.

The Guardians and Scientific Antivivisection

The most ardent antivivisectionists claim that vivisection is not an animal rights issue at all, but rather a problem of scientific and medical fraud. This is the position taken by the Guardians, a recently formed SMO, and by the more established Australian Association for Humane Research. Both groups reject the reformist approach of some of the larger animal welfare SMOs in the United States and the United Kingdom, such as the AHA and the HSUS.

The Guardians describe themselves as "a group exposing vivisection." They were founded in the mid-1990s when they broke away from Animal Liberation Victoria, whose animal rights ideology they believed failed to grasp the essential "evil" of vivisection. According to the Guardians, humans, more so than animals, are the main victims of animal experimentation.

The change of name from Animal Guardians to Guardians and their new logo, which likewise omits the animal, reflect the process of "frame transformation" (Snow et al. 1986) that the SMO has pursued for ideological and strategic reasons. The Guardians now see the ethical issue of cruelty against animals as less effective than the argument against vivisection on scientific grounds, in the same way that the Farm Animal Reform Movement in the United States has replaced an animal cruelty frame with a healthy eating/vegetarian frame.

The Guardians' efforts to enlighten the public are carried out by way of pamphletteering and via its quarterly newsletter, which its vice-president claimed was *"one of the best in the world."* Since the Guardians broke away from the mainstream animal movement, there have been a number of articles in the newsletter that highlight the controversies surrounding animal experimentation within the movement itself. Three of these articles which appeared in the *Guardians Newsletter* provide an insight into the mind of the movement on what many insiders perceive as the most difficult and divisive issue for animal protectionists worldwide.

Although the question mark in the article *Peter Singer—Lab Animal's Friend or Foe?* (*Guardians Newsletter* November 1994) suggests that the jury is still out, the *Dear Peter* letters published in the Guardians' newsletter are unforgiving in their condemnation of Singer's alleged endorsement of provivisection comments on a television program in 1994. In a debate with Dr. Simon Bain of the Australian National University concerning brain experiments on monkeys, Singer is charged with giving these experiments undeserved credibility, allowing Bain to get away with the claim that there are no nonanimal alternatives. Worst of all, it is alleged that "he actually agreed with Dr. Bain's comment that the monkeys experience no physical discomfort." One correspondent representing People Against Vivisection Inc. claimed that "all vivisection has no scientific worth whatsoever" and invoked the names of "genuine" scientists and professionals who "have stated categorically that vivisection costs both human and animal lives, retards medical progress, blocks the path of true knowledge of, and subsequent cures for disease, and is a gross waste of funding, and precious time. This knows **no exception!** [her emphasis]. This is a classic hard-line abolitionist position on animal experimentation that rules out any kind of compromise with vivisectionists or moderates within the animal protection movement, like the AWI, the AHA or the HSUS in America. Singer's utilitarian philosophy and pragmatism will never satisfy the abolitionist. Further reasons for the gap between abolitionists and welfarists are given in the next two stories from the Guardians newsletter.

The article *Animal Rights—The Suicide of a Movement* (*Guardians Newsletter* August 1995) is the work of the awkwardly named Students United Protesting Research on Sentient Subjects (SUPRESS), which is based in California. Its thesis is that the animal rights/welfare movement has prevented itself from defeating vivisectors by clinging to moral and ethical arguments instead of using the available medical and scientific facts that effectively destroy the case for vivisection. The article argues that it is futile to confront the biomedical empire with the emotional appeal of the rights of innocent animals. Put simply, pitting tiny, sick babies against rats, mice, dogs, and cats will always allow vivisectors to win the debate and ensure that the antivivisectionists are labelled antihuman. A similar argument is taken in the last of the three critiques of the mainstream animal protectonist position on animal experimentation.

Again, the author of the polemical piece *Animal Rights, Animal Welfare—The Hijacking of a Movement* (*Guardians Newsletter* Summer 1996), Penny Schultz, takes animal rights/welfare SMOs to task for their complicity in allowing vivisection to continue. She targets the policies of an unnamed Australian SMO for its three Rs strategy—refining, reducing, and replacing animals in experiments—and for failing to include the fourth R, relevance to humans. Also included in the sins of omission is the apparent ignorance on the part of the mainstream animal movement of the widespread challenge of alternative movements—natural and preventive therapies—to Western medicine and Western scientific thinking itself. These movements, the author suggests, coincide with "the new antivivisection movement" and its "revelation" that animal experimentation does not lead to human health.

Of interest here is the process of frame transformation (Snow et al. 1986: 464–81), whereby the original claims of the antivivisectionists that centred on cruelty to animals are replaced by new claims based on scientific, environmental, and human health concerns. With frame transformation, "new values may have to be planted and nurtured, old meanings or understandings jettisoned, and erroneous beliefs of "misframings" reframed (Goffman 1974:308) in order to garner support and secure participants" (Snow et al. 1986:473). In both the article by SUPRESS and the one by Schultz, the perspective transformation is domain-specific rather than global, where a more sweeping transformation takes place. In other words, potential supporters are not being asked to change their beliefs and behaviovrs in several domains, but rather to consider the rejection—in this case of Western scientific and medical practises—of a single interpretive frame. This involves "a reframing of hitherto taken-for-granted aspects of everyday life" (Snow et al. 1986:475) and the substitution of alternative, natural, and preventive therapies. Although domain-specific transformations have the advantage of specificity—for example, an individual can campaign against animal experimentation on scientific and medical grounds but still eat meat—they are not immune to vulnerabilities. In the "new antivivisectionism" for instance, a dramatic breakthrough in cancer research or some similar success directly attributed to animal experimentation could conceivably undermine their case, as experimental medicine had done so effectively in the early twentieth century (Turner 1980:118).

What is more important than these theoretical possibilities, however, are the practical implications of the new antivivisectionism for the animal protection movement. Not withstanding the logic and commitment of the aforementioned protagonists in the debate, their support base appears to be weak, considering the small membership of the two most prominent antivivisectionist SMOs in Australia.

THE FATE OF HUMANE RESEARCH IN AUSTRALIA

That vivisection is a very difficult, divisive issue is a common lament of activists everywhere who are opposed to animal experimentation. In her history of Victorian attitudes toward animals, Ritvo refers to animal experimentation as "the hardest case" (1987:157), while Rodd's bio-ethical reflections on animals lead her to suggest that their use in medical research and the use of pest control as a solution to competition for resources are the only really plausible areas of genuine conflict between human and animal welfare (1992:177). And Elizabeth Ahlston, president of the Australian Association for Humane Research (AAHR), describes vivisection as:

Such a huge issue, you never stop learning. I know it's a very, very nasty subject, but it's also very fascinating. Something is always coming up, more and more evidence to prove how, how crazy it all is. (Interview 1996)

The above adjectives suggest why antivivisectionists find the issue "intractable" or at least "difficult." They claim that although the evidence is all on their side that vivisection is morally and scientifically impoverished, powerful vested interests in the animal research industry ensure that the practise continues.

Animal Atrocities Not Fit to Print

Founded in 1979, the AAHR is one of the oldest antivivisection societies in Australia. As a specialist SMO, its goal is "the total abolition of all invasive experiments on living animals" (undated AAHR flyer). At its peak in the early 1980s, AAHR had 2,450 members; by the middle of the 1990s, the membership had dropped to around 1,000. Ahlston explained that the publicity surrounding the animal rights raid on Dr. Edward Taub's American laboratory in 1981 boosted interest in the SMO:

You see, it was dramatic and it was new in Australia. The media interest was marvellous. And of course, when you have never had dealings with the media, you are naive enough to believe this is going to go on, because there was hardly a letter that I wrote to The Australian, or the Sydney Morning Herald, or whatever which wasn't published. But now it isn't that my letters have gotten worse, they are just as relevant, they won't publish them and Animal Liberation has found exactly the same thing. (Interview 1996)

A recent study of newspaper coverage of anti-animal research activities in the United States between 1983 and 1995 supports Ahlston's observations. The study revealed that publicity peaked in 1986 and then declined steadily into the 1990s (Jones 1996:10). It was in the middle of this period that the AAHR attracted the most attention from the media.

The duration of media coverage of animal welfare issues in Australia is invariably short, and it is usually in response to some dramatic event or stunt. Australia has never had the exposés to rival Silver Spring (Blum 1994) or the Feldberg investigation (McDonald 1994) in the United Kingdom. Antivivisectionists were therefore gratified when a popular teenage magazine, *Dolly*, published an article on animal welfare in 1994. AAHR had an unprecedented number of requests for information following the story's publication. But apart from the occasional feature story in mass-circulation magazines, animal welfare issues rarely appear in the media.

One Australian study that monitored the coverage of environmental issues in prime-time television soap operas found that only 3.7 per cent of the programmes made any reference to the environment. The study noted that the *Neighbours* series, which is popular outside of Australia as well, dealt superficially with the theme of animal experimentation. Using the theme of mice being dissected in the school laboratory, the program presented the polar positions adequately but failed to include any wider discussion of the issues: "Topical discussion of the use of animals for the manufacture of cosmetics was avoided, as was animal experimentation for human medicines" (Rissel & Douglas 1993: 89). A similar dearth of information has been noted in the scientific literature as well.

In the penultimate chapter of her book *In the Name of Science: Issues in Responsible Animal Experimentation*, Barbara Orlans (1993) praises *The New Scientist* in the United Kingdom for running a series of eighteen articles about the rights and wrongs of animal experimentation. She suggests that such a forum is missing in the United States, where there is little criticism in the scientific literature of animal experimentation.[2] Likewise, the animal experimentation issue rarely gets into the public domain in Australia, the last time being in the Senate Inquiry of 1983.

Several activists interviewed for this study said that they enjoyed good relations with the media but believed animal rights issues had had their day. Ahlston commented that journalists are always looking for new stories, and they had moved on to other issues. She also thought that demonstrations were no longer effective, a point made by animal protection SMO leaders in the United Kingdom and Unites States as well. In the case of the AAHR, the grassroots, media-focussed demonstrations of the 1980s, such as "The Great Dog March" in Sydney, have been replaced by a strategy of advocacy via legislative lobbying and education as well as by a subtle shift in emphasis away from the ethics of animal rights to the politics of human health.

Frame Transformation in Antivivisectionism

In their flyer "What we are and what we do," which lists some of the AAHR's main campaigns since 1979, the references are mainly about cruelty and animal abuse—eleven references, all in the 1980s—including the Draize and LD50 tests. Although the first of seven references to health is in 1981, human health issues receive prominence only in the 1990s. Thus, it would seem that the AAHR, like its ideological soul mate the Guardians, has experienced a transformation of its interpretive frame in recent years. Ahlston points this out:

We've always focussed on the human health angle, but I think reading what has happened to the movement, particularly in the United Kingdom, and how it started with the 1876 Cruelty to Animals Act, with only about three hundred to four hundred animal experiments a year and with great Victorians speaking out against it—even Queen Victoria was against it, and Gladstone and Ruskin and Lord Shaftesberry and Cardinal Manning, they all spoke out strongly against it from the ethical and moral point of view—and it made no difference whatsoever. The animal research industry grew and grew and at its peak, I think they were doing about five and a half million experiments a year. And of course, what we are now up against is huge sums of money and jobs for the boys and girls! (Interview 1996)

Ahlston's historical allusions remind us that the humane and antivivisectionist movements of the nineteenth century are the longest-running campaigns in the animal protection movement (Finsen & Finsen 1994). From the outset, both movements were uneasy allies since the focus of the humanitarians in the RSPCA, for instance, was on cruelty to individual animals and not on the institutionalised cruelty of animal experimentation that the antivivisectionists challenged. Yet both movements were united in their opposition to cruelty and in their appeals for compassion toward "the brute creation." And while the leaders of the early Victorian antivivisectionist movement on both sides of the Atlantic argued that vivisection would prolong *human* suffering rather than ease it, the main focus of their concern was the pain and suffering that vivisectors inflicted on animals.

In contrast to their Victorian predecessors who campaigned against cruelty and science, the new antivivisectionists argue that it is futile to appeal to the moral sensibilities of people using the image of the suffering animal. Instead their appeal is to logic: How can experiments on rodents be relevant to the improvement of human health and well-being? To the new antivivisectionist, the issue of animal experimentation turns exclusively on scientific credibility; to the modern-day animal protectionist, it remains the moral issue posed by the spectacle of "the agonized animal." This is the ideological stalemate between the two contemporary movements in Australia and elsewhere that was described above as "the suicide and hijacking of a movement."

According to Ahlston, public opinion in Australia has shifted to the extent that people are now more critical of animal experimentation on scientific grounds. She cites a 1994 Newspoll commissioned by the AAHR that found that almost three-quarters of those sampled (n=1200 nationwide) believed that using animals to test pharmaceutical drugs is misleading because the results can still have harmful effects on humans. Furthermore, two-thirds of the sample felt that there were more effective ways to conduct medical research other than by using animals, more than double the number (31 per cent) who are strongly in favour of the use of animals in medical research (AAHR 1994).

Working with the Devil Himself: Universities, Career Scientists, and Vets

Armed with data of this kind, the AAHR is currently lobbying Parliament to amend the Animal Research Act of 1985 to compel researchers to report to Parliament rather than to each other. The AAHR wants independent inspectors to randomly inspect animal research facilities because, as Ahlston explains:

They inspect labs once every two years, and they give them notice of their coming! I mean, come on now, have you ever heard such nonsense? (Interview 1996)

Ahlston says that the AAHR has good relations with members of Parliament in all of the main parties. However, she believes that Parliament is limited in what it can do. Part of the reason is that responsibility for animal welfare in Australia is a state matter, and yet most of the funding for animal research comes from the federal government.

A more successful strategy, adopted by the AAHR in 1990, was to set up its own trust to fund medical and scientific research that does not use animals. Like other issue entrepreneurs in SMOs, Ahlston started the fund using her own money ($10,000). The New Educational Aids in Medicine and Science (NEAMS) Trust now offers $10,000 annually to fund innovative teaching programs that genuinely reduce or replace the use of animals in teaching.

Targetting veterinarians would seem to be a good strategy, for as Tannenbaum (1993) has argued, veterinarians are in an ideal position to help resolve many of the ethical conflicts between human and animal interests. They are employed in virtually every context in which humans use animals and are generally perceived by the public as humane professionals whose primary interests are animal welfare considerations. There are instances where veterinarians have run afoul of animal protectionists, such as in the recent live export trade in the United Kingdom where the protesters have accused the veterinary profession of complicity in "the cruel trade." But in the main, the vast majority of pet owners—well over half the population in countries like Australia, the United Kingdom and the United States—would generally believe that veterinarians are the best-trained animal welfare professionals available to mediate in conflicts between, say, farmers and animal liberationists. For example, a New Zealand survey (Matthews, Loveridge & Guerin 1994) revealed that veterinarians were seen as the most reliable source of advice on animal welfare for farmers and livestock officers, while 43 per cent of the public favoured veterinarians compared to the 47 per cent that preferred the SPCA.

Animal Experiments as Bad Science, Bad Ethics, and Bad Policy

Ironically, the doctor of philosophy degree (PhD) that universities confer is, in most disciplines, a misnomer since little philosophy, if any, is included in the training. Having a PhD in science does not equip the scientist to make decisions based on the best available ethical approaches to particular problems, such as those concerning genetic engineering. Scientists are, after all, trained to ask predominantly "how," not "why," questions. Or as Komesaroff argues, science is concerned with physical or biological phenomena, the criterion of validity for which is "truth"; ethics, by contrast, is about acts associated with values, the standards of which are not measured by scientific truth but rather by what is "right" or "good" (1992:61–2). While animal protectionists might want the training of scientists to include an ethics component, scientists, it is sometimes argued, would be well within their rights to expect philosophers to engage some of science's questions. But it has to be said that it is the science fraternity that is under scrutiny, not the animal protectionists. As members of the public, animal advocates are increasingly demanding—with the help of freedom of information legislation—the right to know what is being done to animals in the name of science.

Antivivisectionists see universities as among their most intransigent opponents. Universities are infected by a "publish or perish" virus that results in the publication of over 1 million articles in 8,000 biomedical journals each year, according to the authors of a recent viewpoint in *Lancet.* They quote the deputy editor of Journal of the American Medical Association (JAMA), who bluntly states what many animal protectionists fear:

> *There seems to be no study too fragmened, no hypothesis too trivial . . . no design too warped . . . no methodology too bungled . . . no conclusions too trifling . . . for a paper to end up in print.*" (Wiebers, Leaning & White 1994:902–5)

One such paper that the AAHR has criticised concerns the study of obesity in humans. Medical scientists at the University of Adelaide amputated the tail of a small native marsupial, the fat-tailed dunnart, so that the scientists could study how the animal regulates its body composition when it no longer has its tail for storing fat. The AAHR urged its members to write to the vice-chancellor and to *The Adelaide Advertiser* to complain about "the ludicrous experiment." Ahlston also put out a press release and hoped to bring the matter to the attention of the South Australian Parliament by arguing that "it is blatantly obvious that it will not help obese humans . . . and it is outrageous that such experiments receive funding when genuine areas of human need can't" (*The Advertiser* 1997). In response to complaints by the antivivisectionists, the University of Adelaide pointed out that the experiment was funded by the Australian Research Council, which it claimed only funded "significant and high-quality work" (AAHR 1997). Other animal protectionists sent letters to local newspapers, questioning the value of the project and the usefulness of animal ethics committees that they thought were meant to prevent experiments like the fat-tailed dunnart project. Some believed that the experiment was cruel in that it deprived the animal of a tail for the dubious purpose of improving weight control in humans.

The fact that experiments on animals to solve human problems such as obesity and alcohol abuse have been supported by universities and some sections of the scientific fraternity indicates the obstacles confronting antivivisectionists. All their opponents need to do is invoke the claim of "human welfare" to have their experiments—however obscure, irrelevant, or even cruel—approved by the funding authorities. It is for this reason that vivisection is frequently referred to inside the animal movement as *"a very difficult issue,"* and by C S Lewis in the epigraph to this chapter, as a topic on which few can agree. In the next chapter, I look at a number of campaigns concerning the welfare of both domestic and wild animals.

NOTES

1. For animal protectionists, the Draize test and the LD50 are among the most hated tools of the animal researcher who uses live animals to test the safety of cosmetics and household products, like shampoos and shoe polish. In the Draize test, the eyes of conscious rabbits are examined to measure the substance's level of irritancy. The LD50, short for lethal dose, is based on the practise of forcibly giving animals a substance until 50 percent of them die, thus determining the "LD."

2. *Scientific American* has not been an enthusiastic reporter of the politics of animal welfare over the years. A computer search of its articles on animal research uncovered 90 pieces between December 1983 and December 1996, of which only three, all in 1987, dealt with the politics of animal welfare. By comparison, the coverage of animal experimentation issues in *The New Scientist* has been impressive. Elston's content analysis of the magazine between 1970 and 1991 revealed that it had published close to 400 articles on the vivisection debate, in which about two-thirds of the writers were explicitly for or against the use of animals in research; just over half of the total were critical of some aspect of animal experimentation (1992). And of the nearly two dozen editorials, only two attacked animal rights excesses, while the dominant message was to urge scientists to seek constructive dialogue with their critics. The cover story in *Scientific American* in February 1997, therefore, can be seen as a sign of the issue's importance at the end of the twentieth century.

Chapter 10

Street Dogs, Pets, and Dancing Bears: Direct Action/Direct Mail

> Animals are professors of self-esteem: unselfconscious, confident and utterly themselves.
>
> —Gloria Steinem

In this chapter, I focus mainly on companion animals and some of the controversies associated with their protection in the case study countries and in a number of less-developed nations. The work of the World Society for the Protection of Animals (WSPA) highlights the difficulties and dangers of protecting animals in countries where the idea of animal rights is virtually alien. WSPA is a transnational animal protection SMO with offices in London, Boston, and Canberra. It includes both wild and domestic animals within its mission. The single-issue National Canine Defence League (NCDL), based in London, is concerned exclusively with the protection of dogs. The Doris Day Animal League (DDAL) and its much larger counterparts, the Humane Society of the United States (HSUS) and the International Fund for Animal Welfare (IFAW), are all multi-issue organisations based in Washington D.C. In what follows, I look at how these organisations maintain themselves in the animal welfare marketplace.

Animal protection groups compete for funds in an environment where *"the demands on compassion"*—to use the words of English animal activist Mike Huskisson—may have reached their limit. If this is so, animal protectionists have to become much more entrepreneurial if they are to maintain their position in the animal welfare marketplace. For the SMOs featured in this chapter, the politics of direct mail are as important as those of direct action to the success of their campaigns.

COMPANION ANIMALS AND CAPTIVE ANIMALS

The importance of the role of companion animals for the animal movement should not be underestimated. The historian Keith Thomas places great emphasis on the psychological importance of companion animals changing people's attitudes toward nature over the last several centuries, noting the tendency in the early modern period for the rigid boundaries between man and beast to break down (1983:119). Although domestic animals like cattle, pigs, horses, sheep, and poultry were not kept for sentimental reasons, they would ultimately be the beneficiaries of the emotional attachment that people felt toward their pets. Nowadays, most domesticated animals have an animal protection lobby group defending their interests.[1]

It matters little that some of these groups are not much more than letterheads. What is important about them is that they represent organised lobbying on behalf of individual species of animals. The astonishment that the British historian Turner expressed about the number of creatures that have their own pressure groups (1992:318) was noted in the introductory chapter. In the remainder of this chapter, I will focus on a sample of animal protection SMOs that include both domestic and wild animals in their campaigns.

Putting Pets in Their Place

In his pioneering book *Animal Liberation*, the Australian philosopher Singer confesses that he is not particularly fond of dogs, cats, or horses in the way that many people are (1975:viii). Singer thus marks himself as an atypical species of animal protectionist—one who evidently does not have a pet. As noted in the introductory chapter, the keeping of companion animals is one of the distinguishing characteristics of animal protectionists. Singer points out with his customary logic that one does not need to be a pet-keeping animal lover to be involved in liberating animals from oppression and exploitation; according to him, the portrayal of anticruelty protesters as sentimental "animal lovers" has trivialised the treatment of nonhumans as a serious political and moral issue (1990:iii). Singer's view is directly contradicted by Thomas's observation mentioned above (1983). I am making this point simply to show that even the subject of companion animals is not without controversy within the animal movement.

An even greater divergence of opinion on the issue of pets can be found in the views of John Bryant in *Fettered Kingdoms* (1990) and those of the National Canine Defence League (NCDL). Collette Kase of the NCDL's Hope Project *"loves dogs and wants them to have wonderful lives"* (Interview 1996), while Bryant would like to phase pets out of existence. His argument, like Singer's, is unsentimental and unappealing to animal lovers:

> I have not the slightest hesitation in saying that pet animals should be completely phased out of existence. . . . My reason for this view is that pet animals are slaves and prisoners, and I am opposed to both slavery and imprisonment of innocents. (Bryant 1990:9)

Bryant points out that with a dog population in the United Kingdom of about six million, more than half a million young, healthy dogs are killed every year, and about one million strays roam the streets. He argues that under these conditions, the only freedom for the dog is to return it to the wild. Few animal protectionists would agree with Bryant's solution to pet overpopulation.

Bryant's position on companion animals is challenged by Stephen Budiansky. Although not a supporter of the animal or environmental movements, Budiansky argues that "in an evolutionary sense, domesticated animals chose us as much as we chose them" (1992:24). Dogs, for instance, were attracted to human societies for the food, shelter, and protection from the wild that domestication offered. This arrangement suggests that tame animals lead better lives than their counterparts in the wild because being domesticated means that they live longer and that their survival as a species is safeguarded. For these privileges, they pay the price as pets, as laboratorary tools, as objects of amusement, and as food for humans. To Budiansky, this is a fair contract between people and animals. Most pet-keeping animal protectionists would accept the concept of a contract in the case of their companion animals, but not in most other instances of domestication where animals pay the ultimate price for their captivity. In the case of dogs at least, the NCDL believes that there is a sacred contract with "man's best friend."

The NCDL is unique among British animal protection SMOs in its refusal to destroy healthy dogs. The NCDL runs its own network of shelters for strays and attempts to find suitable homes for abandoned dogs. The organisation was established as a charity in 1891 and has Queen Elizabeth II as its patron. NCDL exists to protect and defend all dogs from abuse, cruelty, abandonment, and any form of mistreatment, both in the United Kingdom and abroad.

The charity launched its Hope Project in 1994 when it was discovered that homeless people had thousands of unvaccinated dogs roaming the streets of big cities in the United Kingdom. Apart from offering homeless people the opportunity to have their dogs vaccinated, wormed, and neutered, the Hope Project assists the owners in finding dog-friendly accommodation or, when this is not available, provides advice on re-homing and temporary care for the animal. The Hope Project also assists dog owners who are faced with eviction because of their pets. In its first year of operation, the Hope Project reported that it had not come across any case of neglect and noted the importance of companion animals to homeless people: "The unconditional love and friendship that a dog can provide is invaluable and for those who sleep rough, there is the added benefit of physical warmth at night" (undated NCDL information sheet). As the largest dog-protection group in the world, the defence of canines is its first priority, and the human welfare issue of homelessness taken up by the Hope Project is only a by-product of NCDL's main objective. Even so, its willingness to campaign for homeless people is an effective counter to the claim by critics of the animal movement that it puts animals, including pets, before people and is therefore misanthropic.

PETA's high-profile director, Ingrid Newkirk, does not use the word "pet" because she believes that it is speciesist to do so. And instead of Budiansky's "covenant of the wild," she favours a "more symbiotic relationship" with animals, which she describes as "enjoyment at a distance." Like Bryant, Newkirk believes domestication represents an unnatural relationship with animals that should be phased out:

> We now have burdened ourselves with the custodial obligation to sterilize thousands of animals because we have screwed up their reproductive cycles so much through domestication and inbreeding that they have many more offspring than they normally would. (Newkirk in Hitt 1988)

Yet Budiansky, Bryant, and Newkirk are all in their own way out of step with the mainstream animal protection movement. Since the Industrial Revolution, living with companion animals is how the vast majority of people in the West experience animals. They remain as important in shaping people's attitudes in the late twentieth century as Thomas argued they did in the early modern period (1983). James Serpell sensitively and accurately suggests why it is that animals, in spite of being denied moral status by the neo-Cartesians because they lack language, mean so much to millions of ordinary people:

> Lacking the power of speech, animals cannot participate in conversation or debate, but by the same token, they do not judge us, criticize us, lie to us or betray our trust. Because it is mute and non-judgemental, their affection is seen as sincere, innocent, and without pretence." (Serpell 1986:114)

This is the source of the "bond" that many people experience with their companion animals, and it is also the basis for the reservoir of good will that the animal movement depends on in its campaigns. It should therefore come as no surprise that for many mainstream animal protection SMOs and their supporters, the welfare of cats and dogs is often their most important and abiding concern. In the following case studies, some of the issues concerning companion animals are described from the perspective of a number of multi-issue animal protection SMOs, both big and small.

Lobbying and Litigating at the Doris Day Animal League (DDAL)

The DDAL has been one of the most successful small animal protection lobbies in the United States since the former Hollywood star Doris Day set it up in 1987. In its first five years, the organisation raised over $10 million via direct-mail campaigns to promote opposition in Congress to animal cruelty. Like other successful fund-raising animal protection groups, DDAL has been criticised for the time and money it spends on direct mail rather than on direct action. Even so, as Guither points out, the league is serious about "citizen action" and uses its mailings to promote the public's understanding of animal welfare (1998:56). Some examples follow.

For many people, the pet shop is a source of delight. On closer scrutiny, however, the proverbial cute doggy in the window may have started life in the much harsher world of the puppy mill, where millions of animals are bred for commercial profit. According to the United States Department of Agriculture (USDA), approximately 140 000 dogs and cats were used in animal research in 1993; of these, some 45 per cent came from Class B dealers, 40 per cent were purpose bred by Class A dealers, and the rest came from shelters and pounds (Hellerstein 1994). Class B dealers trade in animals with diverse genetic backgrounds; the animals are bred and raised by others and are sold to researchers for much less than the purpose-bred animals whose genetic background is controlled by the Class A dealers. American animal protectionists have long been campaigning against Class B dealers who, they allege, include stolen pets in their trade and are negligent in following the animal welfare provisions of the USDA. Animal protectionists want Class B dealers prohibited from supplying animals for research. Needless to say, most researchers oppose such a ban, claiming that to use the alternative, purpose-bred animals would be extremely costly and wasteful of animal life; they argue that it is ethically unacceptable for as many as 10 to 16 million animals to be euthanized in United States pounds and shelters while forcing research facilities to produce additional numbers of purpose-bred animals.

From an animal welfare perspective, it is more humane to use purpose-bred animals who have never experienced the freedom of a pound animal, although apart from the American Humane Association, few animal protection SMOs are prepared to say so publicly (Orlans 1993:219). Orlans advocates a compromise solution that involves the continued use of pound animals for short-term, non-survival experiments since, in general, the earlier the end point of death, the better it is for the animal (1993:220).

The clinical language of Orlans' well-meaning compromise would repel antivivisectionists and many mainstream animal protection organisations. SMOs like the Doris Day Animal League (DDAL) and the Humane Society of the United States (HSUS) oppose puppy mills on the ground of cruelty. The DDAL, together with other plaintiffs, has taken the bold step of filing a law suit against the USDA for allegedly failing to stop cruelty in these animal-production facilities. According to the DDAL, the USDA has violated the Animal Welfare Act by not ensuring that dog breeders provide basic humane conditions for the puppies, such as adequate housing, water, food, ventilation, and sanitation. DDAL members were mobilised to send over seventy-five thousand letters and postcards to the USDA, but the government failed to take action on the issue.

The DDAL is primarily a political lobby in Washington D.C. that uses the strategy of petitioning and making presentations to Congress. Its executive director, Holly Hazard, is an attorney who knows how to pressure lawmakers into taking their responsibilities seriously. In the case of the USDA's position on puppy mills, Hazard maintained that the next logical step that would "see an end to government malfeasance that allows unsuspecting consumers to buy sick puppies whose mothers languish in inhumane conditions for their entire lives" (undated DDAL Animal Litigation document). Letter-writing campaigns are among the most common tactics that animal protection activists and advocates use. On the other hand, litigation against individuals or organisations is more risky and requires the kind of financial and professional resources that few animal protection SMOs can muster.

According to Staggenborg, the model for litigation in the social movement sector was the pioneering efforts of civil rights groups who successfully used litigation as a tool for social change (1991:35). Favourable court outcomes have made litigation a popular tactic in other social movements. As I show elsewhere (Munro 1999b), this applies also to various countermovements against animal liberation, which *have* used strategic lawsuits and "McWrits" to stifle the movement's activities.

Some commentators argue that litigation in animal protection has had limited success compared to more aggressive tactics. However, others like Helena Silverstein point out how litigation can provide indirect benefits such as publicity, public awareness, and social movement building (1996). Perhaps the best illustration of this was the McLibel case in the United Kingdom, which has given a new social meaning to the seemingly innocuous hamburger (Munro 1999b). Yet evidence from the ASIS sample of animal protectionists suggests that they perceive filing legal suits and initiating boycotts against businesses involved in cruelty to animals as among the least effective and least justified of tactics. Both the American and Australian respondents agreed that developing animal-awareness education programs and media promotions were more effective in achieving improvements in animal welfare (Munro 1996b). When litigation or boycotts are implemented in conjunction with these other legal activities, success is more likely, as in the case of the HSUS's boycott on puppy mills.

The "Principled Pragmatists" of the Humane Society of the United States

The Humane Society of the United States (HSUS) claims to be the biggest animal protection organisation in the country (HSUS Web Site January 1997) with more than 3.5 million constituents. The SMO covers the key animal rights issues of farm animals, laboratory animals, hunting, companion animals, humane education, bio-ethics, wildlife, and habitat protection.

For at least a decade, the HSUS has campaigned against the inhumane conditions in puppy mills, which supply animals to pet shops throughout the United States. Like the DDAL, the HSUS was frustrated by the USDA's failure to protect puppy-mill puppies. In 1990, the HSUS initiated a nationwide boycott against six of the most notorious puppy mills in the country. Using the slogan "Don't buy puppy-mill puppies!" the HSUS succeeded in mobilising an impressive amount of media coverage, which by the end of the year had forced the closure of many of the state's worst puppy mills. The HSUS stepped up pressure on Congress by calling on its large membership to write to their representatives. They also urged the governor of Kansas to avoid the boycotting of Kansas dogs by taking action against the cruelty of the breeding kennels under her jurisdiction.

According to the editor of *Boycott Quarterly*, boycotts are a form of "economic warfare . . . a hostile act" that should never be entered into lightly (Lyons 1996:4). Lyons points to the futility of a boycott that the target believes will never end since there is no incentive for the target to reform its practises. In the HSUS boycott, the demand was for state authorities to either enforce existing animal protection regulations or introduce new ones covering puppy mills, not to eliminate them all together. There was therefore an incentive for the owners of breeding kennels to provide humane conditions for the animals.

In taking on the vested interests of animal breeders, the hunting lobby, and many other alleged exploiters of animals, the HSUS unleashed a backlash designed to undermine the organisation's integrity with attacks of financial impropriety against the CEO.[2] Even within the animal movement in the United States, the success of the HSUS has created some resentment. I was told by more than one movement insider that the HSUS had become aggressively entrepreneurial and was concerned mainly with building membership and fund-raising. The HSUS was founded in 1954 after breaking away from the more conservative American Humane Association (AHA). But today, critics claim it is no longer the militant advocate of animal rights that it once was, a claim that its media spokesperson would no doubt challenge (see Wayne Pacelle in Chapter 11).

DIRECT-MAIL FUND-RAISING AND THE POLITICS OF EMOTION

People for the Ethical Treatment of Animals (PETA) and HSUS—the big two in United States animal protection—have been both criticised and lauded for their resource mobilisation strategies. Much of the HSUS's literature follows the principle of mobilising emotions and funds by a combination of the obnoxious and the nice. As in most campaigns of this kind, their brochures show endearing pictures of attractive children and adults in the company of equally appealing dogs and cats, with an appeal for a donation. Direct-marketing campaigns have been very successful for the HSUS, which has seen its membership grow from 1 million in 1991 (Jasper & Nelkin 1992:64) to its current constituency of 3.5 million. According to the HSUS's annual report, its total revenue in 1995 was in the region of $38 million, with the bulk ($23 million) coming from contributions and grants. The HSUS is now the largest, and one of the wealthiest, animal protection SMOs in the United States, which ensures that its campaigns are professionally organised and executed.

Direct-mail campaigns are popular with environmentalists and animal welfarists, as flora and fauna issues are colourful and can be promoted in ways that appeal to people's emotions. According to Dowie, direct mail has become the lifeblood of the mainstream environmental movement (1995:44). He points out that direct mail is not just about fund-raising and building membership but is also an educational medium. The same applies to the animal protection lobby as well, which uses its direct mail literature to mobilise, recruit, and educate prospective supporters. Even so, direct marketing can be counterproductive if it is mishandled or is too aggressive.

One of the most blatant examples of this was a letter from the International Fund for Animal Welfare (IFAW) with a dollar coin attached "as proof of the lengths we will go to" to save animals. The letter lists examples of atrocious suffering by animals that IFAW wants to prevent, and it appeals to the addressee to consider sending money "as an investment in the future of the animals." The cleverness in including a real dollar is that it plays on one's conscience—do I keep it, return it, or send a contribution?—even more keenly than do the unsolicited Christmas cards and the like that the same organisation sends to people on its mailing list. IFAW's "Dogs of Despair" campaign consciously seeks to shock people into either sending the SMO a donation or signing a postcard condemning cruelty in the Philippines. Explicit in the IFAW literature is a narrative of cruelty, wherein the suffering of the dogs is described in detail in a four-page letter with accompanying graphic pictures that shock as well as inform the reader. No one who peruses the brochure is left in any doubt about the "facts" of the case as portrayed by the IFAW. The message is that these dogs are just like your own pets. The moral shock that comes with this realisation is designed to induce pet lovers to support the cause financially. In defence of this form of emotional blackmail, it is only fair to note that IFAW is what it says it is, namely a fund for financing animal welfare initiatives.

The Strategy of Direct Action

According to the author of *The Activist's Handbook* (1996), Randy Shaw, to be successful, activist groups need direct-action components that will inspire people to support the cause. As an example of a stunningly successful campaign strategy for animal rights, he singles out PETA for "the willingness of its participants to risk their personal safety, often in undercover, quasi-legal actions, on behalf of animal rights." Shaw also highlights the effective combination of direct action and direct mail at PETA that "uses direct mail as a tool for organizing and mobilizing, not simply to raise funds or to create artificial constituency bases" (1996:233). He claims that PETA has succeeded where comparable environmental groups have failed because the animal rights SMO demonstrates, in practical ways, how its supporters can make a difference to the well-being of animals, for example, by "choosing cruelty-free" products from a list that PETA provides.

The combination of direct mail and direct action is a powerful strategy that organisations like PETA and Greenpeace use to recruit new members and to keep existing members faithful to the cause. People appear more willing to put their money, if not their bodies, at the disposal of direct-action organisations if they believe that their campaigns make a difference. As Greenpeace has regularly demonstrated, media stunts and dramatic rescue operations to save endangered wildlife appeal to people's sensibilities so that giving money or other resources to the cause is understood as a substitute for direct action. Animal protection SMOs discussed in this chapter—NCDL, HSUS, PETA, DDAL, and IFAW—have successfully exploited this sentiment.

A field officer at the World Society of Animal Protection (WSPA) John Joseph suggested that high-profile SMOs are often driven by fund-raising rather than by addressing really challenging animal welfare problems with "hands-on projects in countries where the term animal protection is seldom heard" (Interview 1996). WSPA, more than any other animal protection organisation in this book, sees itself responsible for the animal movement's "too hard basket" and is therefore an appropriate case study on which to conclude this chapter.

WSPA: WHERE ANGELS FEAR TO TREAD

Andrew Dickson, the chief executive officer of WSPA, identifies three issues that he believes are the most important ones for the worldwide animal protection movement: overpopulation of dogs, live animal transportation, and destruction of wildlife habitats (Donald 1998:12). It is the massive number of animals subjected to cruelty in these issues that has inspired WSPA's concern. According to Dickson, WSPA uses a strategy that involves reminding governments of their obligations to outlaw illegal, exploitative practises toward animals. If this fails, documentation by means of undercover investigations, photographs, videos, and the like are used to mobilise public opinion. This typically means that people in more affluent countries become outraged by what they see as unusually cruel practises that they associate with Third World conditions.

According to Carole de Fraga, an animal protectionist associated with WSPA in Australia, the organisation appeals to people because it combines altruism with a sense of adventure (WSPA personal correspondence 1996). In this way, WSPA could not be more different from its sister organisations the RSPCA in the United Kingdom and the MSPCA in the United States. WSPA has over three hundred member societies in about seventy countries, including Australia where it is affiliated with the RSPCA Australia and ANZFAS. Its headquarters are in London where it was originally part of the RSPCA. Today, the RSPCA is WSPA's largest United Kingdom member, although they are separate organisations. However, in the United States, WSPA shares the same premises as the MSPCA, with which it is affiliated.

WSPA is unique as the only animal protection SMO in the world that enjoys consultative status with the United Nations (Garner 1993:190). It can also claim to be unique in its willingness to take on animal welfare issues in less-developed countries where most animal protection SMOs would fear to tread. Of all the organisations profiled in this book, WSPA is the one most likely to have been involved in rescuing starving animals from zoos during the Gulf War in the Middle East, campaigning against bear-baiting in Pakistan, or protesting bullfighting in Latin America and Spain. The organisation has been prominent in dog-rescue campaigns in Bosnia-Herzegovina. They have also shut down illegal cat-killing facilities in Mexico and introduced humane methods of pet-population control in places like India and Colombia. In short, WSPA works with local authorities in countries where there is little or no tradition of animal welfare.

As we have seen, pet protection is one of the animal welfare movement's defining issues. In countries like the United Kingdom, the United States, and Australia, pets are the beneficiaries of most of the existing animal protection laws. In less-developed countries, and in countries like Japan or in parts of Europe, there are virtually no laws safeguarding animals against wanton cruelty. WSPA is at the forefront in campaigning against cultural practices that constitute cruelty to animals from the Western perspective. Needless to say, this often results in hostile confrontation with those whom the SMO has targetted. The Australian representative for WSPA, Carole de Fraga, has described practices such as bullfighting, bear-baiting, and the Southern European habit of keeping dancing bears as "cruelty in the name of culture" (1993:16). De Fraga is careful to avoid any racist allegations by pointing out that it is cultural difference rather than racial characteristics that encourages these forms of cruelty. To emphasise this point, she includes British hare-coursing in the list of cruel practices.

It is instructive to see how a worldwide animal protection society mobilises support both for what the West regards as charismatic megafauna, in the case of wild bears, and for a less-exotic species, like stray dogs. Both are worldwide problems for animal protectionists and are the focus of two of WSPA's best-known campaigns, the "Libearty" campaign and the Pet Respect program. To be effective, the SMO needs to weave its way through a cultural minefield in countries where animal rights is an alien concept.

India, a country with millions of vegetarians and home to the philosophy of Ahisma (nonviolence) is also home to the illegal practice of dancing bears. Although bear dancing was banned in India in 1991, Mandaris or "bear people" continue to procure bears for the purpose of entertaining tourists for money. WSPA reports on a case where a state minister for the environment tried to enforce the law against bear dancing. This led to a wave of public sympathy for the Mandari and a mass demonstration by bear trainers that combined to intimidate the authorities in New Delhi from taking further action against the practise. In conjunction with the Animal Welfare Board of India, WSPA is currently funding a research survey designed to find out more about the bears and their keepers. So far, the results are promising and suggest, among other things, that many gypsies would cease using dancing bears if they were given alternative jobs (WSPA personal correspondence 1997).

A strategy that is more immediately beneficial to the animals is the concept of a bear sanctuary. In Greece, for example, a wine company has provided some twenty-five acres that WSPA hopes to use as a refuge for performing bears; a similar sanctuary is planned for Turkey. The idea of buying land to be used as a refuge for animals is one that was borrowed from environmentalists and conservationists, such as the wealthy Nature Conservancy Fund in the United States, the British Woodland Trust, and the recently formed Australian Bush Heritage Fund. These groups have all recognised that one of the surest ways to prevent environmental degradation to particular parcels of land is to buy the land and preserve it as a trust.

This is a good example of strategy spillover between different social movements, although only national or transnational SMOs that have a substantial membership and financial basis, such as the League Against Cruel Sports or WSPA, can afford to embark on property acquistion on this scale. While the use of sanctuaries for the benefit of captive animals is economically feasible even in remote areas of northern India, the same cannot be said for the millions of suffering stray animals in India's huge cities and towns. The scale of pet overpopulation in India, as discussed in the following section, is one of WSPA's most daunting challenges.

Pet Respect Campaign

In 1994, WSPA began it's global Pet Respect campaign to establish humane and effective pet-population control in developing countries and to teach people respect for cats and dogs. I interviewed some of WSPA's field officers and administrative staff in the United Kingdom the United States, and in Australia, all of whom had worked in countries that generally have no animal protection laws and that lack welfare measures, such as licensing, animal vaccination, and spay programs. In encouraging spay/neuter programs, WSPA uses the three-pronged approach of education, sterilisations, and leglislation, the latter focussing on breeding, anticruelty, and registration measures that, in most developed countries, are well accepted in principle, if not in practise.

The cultural difficulties associated with the Pet Respect campaign are illustrated in the case of India, a country with one of the world's largest stray dog populations and a nation famous for the largest number of stray cows in the world. While Section 48 of the Indian Constitution protects the cow, an animal revered by millions of Hindus, there is no such protection for the millions of stray dogs in the streets of India. Other animals suffer a similar fate due to contradictory laws in the country. Cultural and religious factors mean that the "sacred cow" gets a better deal than the domestic canine or animal destined for Indian dinner tables does. In 1994, WSPA joined forces with government and private organisations in Jaipur, in the north of India, to initiate the country's first animal birth-control program. One of the cooperating SMOs was Help in Suffering (HIS), with which Christine Townend, the Australian activist and founder of the first Animal Liberation branch in Australia, has been working for the past several years (Townend 1995).

WSPA needs to tread a cautious path in its animal welfare work in a country where the cultural tradition of Ahisma means that there is a strong aversion to the taking of an animal's life. According to an HIS officer, cats and dogs are often in an advanced state of malnourishment when they are brought to shelters since people decline to feed the animals meat on religious grounds or to kill an animal humanely if it is suffering. A further cultural complication is that because of religious beliefs, there has been resistance to the introduction of the prestunning of animals since many vegetarians are opposed to the modernisation of Indian slaughterhouses for fear that this will lead to an increase in meat eating (Townend 1993:12).

WSPA field officer John Joseph explained to me that the Pet Respect dog-control campaign is probably one of the biggest issues in the Third World. Joseph has been involved in projects in India where WSPA has gone into municipal dog pounds and seen the slaughter of as many as 100 hundred to 200 hundred dogs by electrocution or strychnine poisoning per day. He claims that this mass, inhumane euthanasia has been going on for the past three or four decades without reducing the dog population by a single animal. WSPA has succeeded in convincing the Indian government that its mass euthanasia solution solves nothing at all. The authorities are just beginning to get some very promising results by stabilising and controlling dog populations through spaying and neutering. Joseph pointed out that this has been done on the back of a rabies vaccination program that has human, as well as animal, welfare benefits. Overcoming cultural barriers to achieve Western standards of animal welfare is no easy matter, and yet with its Pet Respect campaign in India, WSPA has made significant progress.

Another example of WSPA's where-angels-fear-to-tread philosophy is its Pet Respect campaign in Bogota, Colombia. As a country with a reputation as one of the most lawless places in South America, Colombia represents one of WSPA's most formidable animal welfare challenges. In an article entitled "No Mountain Too High" (1997), the program's coordinator, Stephanie Roderick, explained how WSPA's Pet Respect campaign succeeded in getting the Colombian government to stop the inhumane practice of electrocuting 300 to 500 stray dogs every week. Working in conjunction with the sole animal shelter in Bogota, representatives of the zoonosis center, and veterinary professors and students, WSPA utilised the services of animal control experts from the MSPCA to show how veterinarians and shelter workers can initiate and implement education and sterilisation programs to save dogs' lives. Affordable alternatives to electrocution were demonstrated, and equipment and state-of-the-art veterinary supplies were donated to the local zoonosis center. According to Roderick, these steps represented the successful beginning of the Pet Respect campaign's three-tiered strategy of legislation, humane education, and sterilisation in Bogota.

The "Libearty" Campaign

Work of this kind is costly. WSPA, like its partner the RSPCA, expends much of its resources on hands-on programs such as Pet Respect. Its relationship with the MSPCA in the United States, for example, provides it with a solid donor base. WSPA staff members told me that campaigns were usually organised if they had public appeal. John Joseph at the organisation's London headquarters noted that animal protection organisations were now competing for funds, and therefore WSPA needed to be professional in its operations. This can mean an emphasis on fund-raising at the expense of hands-on projects or an approach that seeks to balance traditional animal welfare work and fund-raising. Joseph claims that the organisation, although run like a business, is not market driven. Quite the reverse:

> *It's a form of business like any other, but I think the important thing for us is that we don't lose sight of the goal of why we are here. I think, in animal organisations, we have very strong ethics in terms of why we are doing something . . . and ensure that everything that we do in terms of marketing is driven by field projects and not the other way around. I think that's when the problems occur, when marketing forces drive the projects.* (Interview 1996)

Interestingly, WSPA is one of the only organisations I studied that actually went so far as to offer training tips to its supporters in the techniques of fund-raising and sponsorship. In one of its *Libearty News* magazines (WSPA July 1993:5), for example, it devoted half of the sixteen pages to providing readers with fund-raising ideas. This suggests, contrary to John Joseph's view, that WSPA may be no different than the other animal protection SMOs discussed in this chapter, in that it has to compete energetically for money to finance its costly operations.

As an organisation that competes for funding in the animal protection marketplace, WSPA needs to ensure that its campaigns resonate with its financial backers. WSPA is as adept in the politics of emotion as the other successful animal SMOs discussed in this chapter are. Its glossy magazine *Libearty News* is replete with dramatic, contrasting pictures of beautiful, healthy bears in the wild that the organisation has rescued from "the living hell of bear farms in China and Vietnam." Other bears rescued from bear-baiting in Pakistan or from the banquet tables of Thailand are given their own names—Eric, Paddington, Black Cloud—to bring their individuality home to readers. These bears are contrasted with the maltreated, pathetic creatures that are held captive in bear farms, paraded as dancing bears, exhibited in zoos, or strung up as trophies.

The hunting of bears has been described by Vicki Kysar, an Australian animal protectionist working in WSPA's Boston office, as "the shameful hunt" because it involves the infliction of cruelty without any obvious risk to the "hunter" (1996:22–23). These hunts include baiting the bear, by which the hungry animal is tricked into taking a bait and then is shot at close range by the shooter, and hounding the bear, in which dogs fitted with radio collars are used to run down the bear for the hunter to shoot.

WSPA's "Libearty" campaign is directed at liberating bears from suffering in all of the contexts in which animal liberationists campaign for other species of animals. Bears are dissected for their body parts, which are used for medicinal purposes. They are used for entertainment and sport in hunting, in bear baiting contests, or as dancing bears. And finally, bears are eaten in ways that would shock the average carnivore: "The live cub is slowly lowered into a tank of hot water while the guests watch. They eat its paws and meat, then take the gall bladder home to dry" (undated "Libearty" promotional brochure). This latter practice is said to be enjoyed by tourists from Korea and Taiwan, who travel to Thailand for the experience.

CULTURE, ANIMAL WELFARE, AND STRATEGIC CONFRONTATION

These atrocities inflicted upon bears reinforce Roger Scruton's view that the most important question about our relations with animals is not *whether* we should experiment upon them, hunt them, or eat them, but *when* and *how* (1996: 8). According to Scruton, the concept of virtue helps us understand how we should treat animals. The treatment of bears in these contexts, which is not restricted to non-Western societies, "compels us to distinguish virtuous and vicious attitudes towards other creatures, regardless of whether those creatures are moral beings like us" (Scruton 1996:55). In all three instances outlined above in WSPA's brochure in which humans use bears, there is a level of cruelty that exceeds what most Westerners—specifically the Americans, Britons, and Australians referred to in this book—would regard as acceptable. Scruton, a critic of hard-line animal rights, describes these ways of treating animals as "the pornography of pain" in which "suffering becomes an object of interest for its own sake" (1996:107).

Some readers will no doubt object that a comparison between WSPA's exposé of the vicious treatment of bears—mainly, though not exclusively in Asian countries—does not excuse the cruel treatment of farm, research, captive, and companion animals by Westerners as described throughout this book. For example, foxhunting, which is described in the next chapter, is a bitterly controversial blood sport that its many opponents would see as every bit as vicious as bear-baiting. The difference is that bear-baiting and the other forms of bear harassment that WSPA denounces are examples of "gratuitous harm," to use a term by Wayne Pacelle of HSUS. Scruton, who is at the opposite end of the animal rights debate, agrees. But unlike Pacelle and other animal liberationists, he argues that on environmental, pragmatic, and utilitarian grounds, foxhunting is morally permissable. He claims that whatever vices might be detected in the hunting field, sadism is not one of them. This, however, is not the case in the hounding, baiting, and persecution associated with the hunting of bears that WSPA is seeking to outlaw. Ironically, hunting with hounds is likely to be outlawed long before these more cruel practices outside the United Kingdom are.

Regan takes a different view concerning cultural differences in the treatment of animals. One of the tactics he advocates might be called "strategic confrontation." He explains:

I think the thing that I would do over and over again [as a strategy] is to show people how in Korea, in China, and so on, "pets" (so-called) are chosen, thrown in boiling water, skinned alive, thrown in vats, drowned, then cooked. And then I would show them what happens to hogs at slaughter. I think the connection just stares you in the face. The only thing that's different is that in Korea and China, they're more honest about what they do. It's more public. In the USA, and other so called "advanced" nations, it's hidden behind closed doors. (Interview 1997)

There is much to be said for Regan's tactic here, although "strategic confrontation," as I call it, may be susceptible to the same kind of reaction that deprived people show when confronted with the suffering of others. Such people are usually not consoled by comparing their disadvantaged lives with, say, the misery of a starving Ethiopian. In short, poor Americans judge their predicament by looking to their well-off neighbours rather than to examples of poverty in the Third World. But this reaction occurs essentially among fellow human victims, and it is quite likely that Regan's tactic might work when the victims are animals. A pig that suffers during its slaughter in Oklahoma might well elicit the same compassion as the bear cub boiled alive in Thailand does. More likely, however, is that Westerners addicted to their pork sausages and ham sandwiches would prefer not to see behind the closed doors of the slaughterhouse and would prefer to agitate against the barbarity of eating bears, dogs, or horses that are not normally part of the Western diet. As a tactic, it would make more sense to omit the cultural comparison and expose people to the conditions of battery farms and slaughterhouses in their own backyard, so to speak. This should be enough to turn stomachs, if not hearts and minds.

WSPA has, however, tried to make effective use of a form of strategic confrontation in its antibullfighting campaigns. Like many of its other campaigns that are consigned to the "too hard basket" by other animal protection SMOs, the issue of bullfighting raises complex cultural questions as well as hackles, since opposition to it threatens strong traditions and cultural values. For instance, a WSPA poster features a dark-eyed, white-skinned young man with full lips and a fleshy nose whose bodily integrity is painfully destroyed by a large bull ring that pierces the nose. The caption above the poster reads, "Bullfights would stop if tourists weren't **led** there" [emphasis added]. Underneath, there is reference to bulls being "drugged, tortured and stabbed to death." The tourist's face in the poster has been rendered therianthropic, producing a disturbing image of "a human treated *like an animal*" (Baker 1993:224). For most sensitive people, the image of the man with a bull ring would surely elicit feelings of pity for the unpictured bulls, and if it did, the poster has served its purpose. Nonetheless, Baker reveals that a number of British airports with flights to Spain banned the poster because of its disturbing message, and so it is not known how many tourists were turned off Spain as a consequence. Nor is it known how many prospective supporters are turned away from the animal movement by graphic photos of cruelty to animals.

What we can say is that the direct-action and direct-mail tactics of the animal protection SMOs described in this and the previous chapter all involve the use of strategic confrontation that affects people in different ways. These include tactics ranging from IFAW's explicit use of the politics of emotion in its graphic atrocity stories to the boycotts and strategic lawsuits of HSUS and DDAL.

Animal activists and advocates need to tread a cautious tactical path between being too timid, on the one hand, and *"breaking down the laboratory door,"* on the other, if they are to tap the reservoir of goodwill that Western publics have for animals. IFAW, as I have suggested in this chapter, runs the risk of alienating support for the movement by its aggressive methods of "selling compassion." The politics of emotion used in direct-mail fund-raising, like different forms of strategic confrontation involving cultural values and tradition, have to be carefully managed if they are to avoid alienating potential supporters. The World Society for the Protection of Animals, I believe, has been among the most effective animal protection SMOs described in this book in achieving the right balance between the tame-cat and the outrageous. Its "Libearty" and Pet Respect campaigns are good examples of the effectiveness of strategic confrontation, not unlike the reintegrative shaming strategy that was so successful in Spira's campaigns, referred to in Chapter 9. And like Spira, WSPA does attempt to encourage alternatives to the practises that it seeks to ban, such as the retraining of gypsies dependent on their dancing bears. In many of its campaigns, it has worked closely with the local authorities to find a satisfactory outcome for both the animals and their human (ab)users. Where this has not been possible, for instance in the Spanish bullfighting issue, the more conventional tactics of the animal movement—confrontation, emotional politics, and moral shocks in the context of direct action and direct mail—are used with varying degrees of success. In the next chapter, I look at what many animal liberationists at century's end believe is their most promising campaign.

NOTES

1. A recent directory of animal advocacy organisations in the United States includes the following groups that have been set up to defend individual species of domestic and wild animals: Alley Cat Allies, American Society for the Welfare of Cats Inc., American Tortoise Rescue, The Bear Alliance, Buddy Dog Humane Society Inc., Concern for Critters, Dolphin Freedom Foundation, Greyhound Friends Inc., The Home for Unwanted and Abandoned Guinea Pigs, House Rabbit Resource Network, Opossum Society of the United States, and PIGS—A Sanctuary (*The Animals' Agenda* July–August, 1996). Similar groups exist in the United Kingdom and Australia. An obvious omission in this impressive list of animal protectors is an organisation for the welfare of insects, which account for two-thirds of all species.

2. A front-page story in *The Washington Post* (Thompson 1996:1) reported on allegations by an ex-staffer alleging financial impropriety at the Humane Society. The article showed photographs of apparent perks of office that could be damaging to an organisation that advocates the ethical and humane treatment of animals. *The Washington Post*'s photograph pictured a large house described as the residence of the Humane Society's CEO, who lived there for six years rent-free. The text of the long article reports that the 1995 salary of the CEO was $237,871, with the HSUS president earning a yearly salary in excess of $200,000. But most of the article is about the financial affairs of the former staff member and the society's lawsuit against him. Stories alleging financial scandals in the animal welfare industry and comfortable lifestyles of senior executives must inevitably harm the organisations, not least because animal protection advocacy is upstaged by the distraction of legal and public relations work that has the potential to be a financial and emotional drain on the SMOs involved.

Chapter 11

Hunting the Hunters: The Winnable Issue

The fox knows many things, but the hedgehog knows only one big thing.
—Archilochus

The enigmatic quote above neatly sums up the difference between social movement tactics and strategies. It is the hedgehog who knows the strategy or the "broad organizing plans" (Turner & Killian 1987:286) of the movement and the fox who is expert in the specific techniques that movements use to implement their goals. The animal protection movement, like other conscience movements—for example, those that represent refugees, children, or political prisoners—needs the services of both hedgehogs and foxes to be effective. Furthermore, conscience movements are more dependent on the goodwill of the public than beneficiary-based movements, since the latter draw upon an existing infrastructure of support that groups with weak or nonexistent citizenship status lack (Zald 1988). The organisational strategies and tactical mechanisms that the animal movement employees must therefore be constructed in ways that will appeal to, rather than alienate, potential supporters and adherents. In this chapter, I describe the strategies and tactical repertoires used in one of the movement's main campaigns. The campaign against blood sports is, after antivivisection, the movement's longest-running issue. Ironically, it is widely believed to be the most winnable issue for the movement at century's end.

ENGLISH, AMERICAN, AND AUSTRALIAN HUNTING PRACTISES

The single-issue League Against Cruel Sports (LACS) has been campaigning in Britain on behalf of wild animals since 1924, while a number of animal protection SMOs in the United States and Australia have taken up the cause of wild animals in more recent times. Hunting is one of the most contentious animal issues in the United Kingdom since the protagonists on both sides believe that their cause is just. For many conservationists, hunting is an eco-friendly and traditional method of culling wild animals that they claim is more humane than the alternatives of mass shooting or snaring are. Animal rights opponents argue that killing wild animals, for whatever reason, is a blood sport and an exercise in brutality that results in unnecessary pain and suffering for innocent animals.

In the United States, the issue is no less emotional, although the traditions are quite different. Canned hunting, for example, is a recent example of the commercial exploitation of animals that allows hunters to select, for a price, any wild animal for target practise. The HSUS believes that almost any animal is used—bears, boars, buffalos, or more exotic beasts such as the rhinoceros—if the hunter is prepared to pay. Shooting, not hunting, is the more accurate term for this practise, which does not require any true hunting skills on the part of the shooter. And although these canned hunts are a world apart from the traditional English sport of hunting on horseback, they share an exclusiveness that only the relatively rich can afford.

Hunting in Australia is also different. Apart from the killing of waterbirds and kangaroos, hunting is not, by comparison, an emotional issue. Nor is the antihunting lobby large by comparison to the English and American lobbies. Franklin (1996) has explained why antihunting and anti-angling activism in Australia is not as intense as it is in the United States and the UK.[1] Even so, the campaign against duck shooting has been described as the most action-packed, hands-on animal rights campaign in Australia and Laurie Levy, the director of the Coalition Against Duck Shooting (CADS), is one of the best-known animal liberationists in the country.

There are undoubtedly real differences among the attitudes of the public to hunting in the three case study countries, although not for animal rights/welfare supporters who do not distinguish so readily between exotic and indigenous, or game and nongame, animals. In the ASIS sample of American and Australian animal protectionists, for example, hunting wild animals with guns ranked fifth highest in moral reprehensibility from a list of fifteen human practises involving animals. It can safely be assumed that animal rights/welfare supporters in the United Kingdom, the United States, and Australia are no different in their opposition to hunting, whether it is foxhunting, canned hunting, or duck shooting, if the crucial ingredient of cruelty to individual animals is involved. Animal protectionists in the relevant countries would want to use the suffering of individual animals as the most salient criterion for determining the necessity of killing the fox, the captive deer, the wild feral pig, or the waterbird.

Hunters are able to cite numerous reasons to justify their pursuit of wild animals: hunting is good for the animal since it keeps them alert and therefore less vulnerable to harm by humans; hunting is good for the countryside as culling keeps animal populations under control, thereby reducing adverse impacts on the environment; hunting is good for the local community and economy since it brings in revenue through the purchase of goods and services. Social scientists, such as the sociologist Jan E. Dizard (1994) and the psychologist James A. Swan (1995) have made such claims to promote the virtues of hunting and to defend it against attack. Devotees of the hunt have also made more esoteric claims in support of hunting (Ortega y Gasset 1972), while others such as Cartmill have allowed hunters to speak for themselves but have found their motives generally wanting (1993). Cartmill's meticulously researched study of hunting through the ages finds little justification for hunting and none at all for sport hunting. This is the position taken by the major antihunting organisations, LACS, HSUS, and CADS, and their leaders, who are profiled below.

LAURIE LEVY: DUCK PROTECTOR-GENERAL OF VICTORIA

In ancient times, ducks were called "prophets of the wind." Today, the description is an appropriate metaphor for the Coalition Against Duck Shooting (CADS), as a harbinger of change in the way that humans treat wildlife. The director of CADS, Laurie Levy, is a former television cameraman whose experience in the media industry has helped make the coalition's campaign one of the most high-profile media stories in Australia (Munro 1997a; 1997b). Levy, who is affectionately known as Victoria's Duck Protector-General, has been a full-time, unpaid animal activist for the past decade. After the success of the "save the whale" campaign in the 1970s, he took up the cause of duck rescue in the Australian state of Victoria, where duck shooting attracted some ninety thousand shooters annually. Initially, Levy asked the state government to change the opening time of the duck season, which effectively meant a reduction of one hour. He claims that he then had the whole shooting community and the government up in arms against him (Interview 1994). This incident led to the rise of CADS as one of the most successful animal rights groups in Australia. After ten years of campaigning by CADS, the number of duck shooters has declined to about a quarter of their original numbers, largely as a consequence of the coalition's success in framing the protest as a duck-rescue operation. I have explained elsewhere (Munro 1997a; 1997b) how CADS has achieved this outcome with the help of the electronic media. The metaphorical "Duck (liberationist) Shoots Man" headline has much more appeal than the conventional "Man Shoots Duck" story.

Despite television's framing of the campaign largely as a narrative of protest and its distortion of the protesters' grievances, the graphic images of "Duck Wars"—"the war on the wetlands," "the slaughter of innocents," and "the carnage of this cruel macho sport"—have ensured the public's sympathy for the animals as "sentient subjects who feel the world." Levy knows how to use these images to arouse the emotions of the public and to prod them into supporting the coalition's cause. In one television program, Levy is shown nursing an injured bird. In exasperation, he addresses the audience:

> We brought out a record number of wounded birds, birds that have been shot through the back of the eye, through the back of the head, through the wings. And here's a young signet, and it's been shot through the neck, and this is a magnificent, beautiful, young bird. You know, duck shooting, the brutality that we see out there every year, is just unacceptable. Duck shooting is not sport. It's cowardly, it's violent, and it's antisocial. And that's why duck-shooting numbers are dropping so dramatically. (GTV 9 *Today* 21 March 1994)

When asked to explain his activism, Levy says that he has always tried to stand up for the underdog:

> *I've always hated injustices. I've always hated to see the powerful and the strong exploit or kick around the weak, whether this be humans or another species . . . It's purely from an injustice point of view. . . . I see other species as having rights, especially the right not to be abused or harmed or to suffer, especially for humans to make a profit out of it or to get fun from it.* (Interview 1994)

These are sentiments commonly expressed by many of the *compassionate beasts* discussed thus far, as well as by Levy's counterparts at HSUS and LACS, who are profiled next.

WAYNE PACELLE: NEW BREED OF PROFESSIONAL ANIMAL ADVOCATE

Although the postal address for the HSUS is downtown Washington D.C., its two hundred or so employees are housed in a large complex in Gaithersburg, on the outskirts of the capitol, where rents are lower. It was here that I met the animal rights/vegetarian advocate Wayne Pacelle, the HSUS's vice president for government affairs and media. In his early thirties, Pacelle is one of the leaders of the new breed of animal protectionist in the United States. Pacelle finds inspiration in the work of early animal protection pioneers like Henry Bergh, the founder of the American Society for the Prevention of Cruelty to Animals, and in the modern ethical teaching of Singer's *Animal Liberation* (1975). As a persuasive and articulate advocate, he is eager to promote the cause of animals in both the electronic and print media. His name appears more often than most American animal protectionists in the general literature on the movement. The hunting devotee Ted Kerasote, for instance, described a day in Pacelle's life in the penultimate chapter of his book *Bloodties: Nature, Culture and the Hunt*, revealing among other things his subject's motivations:

> At baseline I'm against cruelty. . . . The reason that I campaign on the issue of animal rights is that I intensely dislike suffering, and the fact that humans unnecessarily perpetrate suffering on animals. It's more of an intellectual/philosophical motivation than it is a hands-on one. (1993:250–70)

Like his philosophical mentor Singer, Pacelle is not especially fond of animals. In the same way that he would oppose slavery or apartheid without any firsthand experience of these systems, he objects to cruelty and suffering inflicted on animals as morally wrong. Kerasote's observation of Pacelle reveals that he is not averse to inflicting verbal wounds on women in fur coats, something he says that he does not enjoy doing. As he explained to Kerasote:

> One has to have an element of direct action and confrontation. And I think that the fur issue is one that requires some activity on the streets. I don't use spray paint. I don't consider it traumatic or terroristic to tell someone what they're doing is cruel. (1993: 265)

As one of the HSUS's vice presidents, Pacelle sees himself as an activist working for an organisation that he describes as made up of "principled pragmatists," in that

> *We have a core set of values that are not as ideologically pure as some other groups, such as PETA, which I would consider to be to the left of us.* (Interview 1996)

He explains how the HSUS works within the system and eschews doing anything that is illegal. Pacelle suggests that the SMO's large membership can be attributed to the Humane Society's respectable image and reputation as an organisation that *"doesn't sound like it's trying to break down a laboratory door."* It has also targetted animal lovers by focussing on companion animal issues for most of its history, although it does deal with all the seminal animal rights campaigns as well. According to Pacelle, *"the HSUS's global mission is to protect animals from human harm, to preserve their habitat, and to foster respect, compassion, and responsibility."* He explains that the HSUS does form alliances with like-minded animal protection SMOs when it can, but is *"strong enough and big enough to proceed without alliances."*

Pacelle himself oversees a staff of twenty in his role as the principal national spokesperson for the HSUS in the United States, a job he has had since early 1994. Previously, he was the national director for the Fund For Animals, was the founder of the Washington Wildlife Alliance, and was the associate editor of *The Animals' Agenda*, the main organ of the movement in the United States.

In 1996, Pacelle was named "rising star in politics" by *Campaigns and Election Magazine*, no doubt because of his political and media skills. As the founder of a statewide ballot initiative committee, he is responsible for establishing and implementing the HSUS's legislative program in Congress and in the states via the ballot initiative process. Before joining the HSUS as one of its several vice presidents, Pacelle established a political action committee to finance a statewide ballot initiative to ban the trophy hunting of bears and cougars with dogs or bait: "I intend to make candidates responsive to humane concerns. It would not be an overstatement to say that the animal movement is considered impotent by some political observers. Our numbers have simply not translated into political power." (*Animals' Agenda* undated: 14, 6, 28–30).

Opposition to recreational hunting is close to Pacelle's heart. As a college student at Yale University, he got involved in some of the early antihunting protests in the United States and took part in disrupting hunts. He is now involved in attempts to ban the use of cruel traps and to prohibit the hounding of bears and bobcats by the hunting fraternity. Inevitably, this has brought the HSUS into confrontation with one of the most successful lobby groups in the United States, the National Rifle Association (NRA). As in the United Kingdom, hunting is an emotion-laden issue and one that Pacelle believes will dominate the animal welfare legislative program for years to come: "There are going to be more measures relating to hunting and trapping than any other social issue, more than gay rights, taxes, anything" (*The Daily Telegraph* 27 July 1996). Yet it is not an exaggeration to say that American public opinion, as in the United Kingdom, is against most forms of hunting. Pacelle echoes almost the exact words of Mike Huskisson when he describes why most people are vehemently opposed to cruelty to animals:

> *Humans do have the capacity for empathy, and we can see the suffering that animals endure. There's not anything deeply intellectual about a response to somebody hitting a dog over the head with a bat. It is a reflexive emotional response to protect the dog, just as one would protect a child or protect another creature that could not defend himself or herself.* (Interview 1996)

Pacelle argues that there are some practices, especially associated with hunting and trapping, that are repugnant and out of step with public sentiment. The HSUS campaigns against the worst abuses, like cruel traps or inhumane puppy mills, by attempting to codify prevailing public sentiment on these issues. It is a long and painstaking process involving, in the case of ballot initiatives, the gathering of the many thousands of signatures that are needed to get a measure on the ballot for a full vote by the public. Writing to the HSUS membership, Pacelle points to Proposition 197, the first statewide ballot measure designed to reverse a voter-approved wildlife protection measure in California. In March 1996, voters defeated Proposition 197, which would have allowed the trophy hunting of mountain lions on the pretext that the animals posed a threat to humans. In conjunction with the Sacramento Mountain Lion Foundation, the HSUS succeeded in countering the hunting and trapping interests who had poured half a million dollars into promoting the measure. Pacelle believes that citizens' groups will be more active than ever in the next few years in opposing cruel and unsporting hunting and trapping practices (Pacelle 1996:6, 8).

The League Against Cruel Sports in the United Kingdom, like the HSUS, sees a new wave of public sentiment against the slaughter of wildlife that puts hunting and trapping in the winnable category, as opposed to the more intractable issues of vivisection and factory farming. The English practise of hunting with hounds is the most traditional and also the longest-running target of the antihunting lobby, which is made up of both animal rights activists and animal welfare advocates in the United Kingdom. Before I profile two of its leading advocates and activists, some recent background material from both sides of the issue is described in the following section.

HUNTING WITH HOUNDS AND THE LEAGUE AGAINST CRUEL SPORTS

In recent years, one of the policies of the League Against Cruel Sports (LACS) has been to buy large tracts of land in hunting territory, often surreptitiously, that they designate as animal sanctuaries. Animals pursued by hunters can find safe refuge if they happen to enter the sanctuaries during a chase. Evidently, this has saved many animals who have been inadvertently chased onto the sanctuaries.

In the summer of 1996, I was invited to attend an open-day weekend sponsored by the league at one of its animal sanctuaries in Somerset. It was held at one of the recently acquired sanctuaries, deep in the heart of traditional deer-hunting territory. During the weekend, I spoke to dozens of the league's supporters and staff, including people representing other animal protection SMOs that ranged from the RSPCA to the Hunt Saboteurs Association (HSA) and even the Animal Liberation Front (ALF). In understanding the activities of the league, there were two individuals who proved most useful. John Bryant was, at the time of the interview, the league's press and research officer and has since become its joint chief officer. Mike Huskisson is a consultant to the league and is founder of the Animal Cruelty Investigation Group (ACIG), which since it commenced its undercover operations in 1989, has been responsible for some major exposés of cruelty. Bryant and Huskisson are a formidable duo of organisational advocate and grassroots activist who have demonstrated a unique combination of legal, nonviolent direct-action and legislative lobbying that many believe will result in a ban on hunting with hounds in Britain by the turn of the century. A brief account of the background of these two animal protectionists provides an insight into the work of animal welfare advocacy in Bryant's case, and of animal rights activism in Huskisson's.

JOHN BRYANT: PASSIONATE ANIMAL WELFARE ADVOCATE

John Bryant's involvement with the animal movement goes back to the early 1970s. He was elected to the RSPCA national council in 1972, at a time when the venerable society was in the throes of a bitter controversy over its lukewarm stand on blood sports, which the anticruelty SMO had defended since the 1960s (Sheppard 1979). The hunting fraternity had successfully gained membership of the RSPCA, thereby protecting its activities from the society's condemnation. Bryant was one of the 25 radicals in the Reform Group who sought to compel the RSPCA to act against blood sports. In 1975, the Reform Group disbanded, having achieved its objective of getting a majority of sympathisers on the RSPCA's council. In *Fettered Kingdoms* (1990), Bryant observes how the RSPCA has changed fundamentally as an animal protection SMO. He notes that although it took until 1975 for it to actually condemn foxhunting, the society now has a clearly stated antihunting policy upon which it is prepared to act. Furthermore, more militantly outspoken animal rights SMOs like Animal Aid—on whose council Bryant served as chairman for four years—are no longer viewed by the RSPCA as suspect.

Animal rights is a religion to Bryant, not a political campaign, although he admits strict devotion is always difficult. As joint chief officer of the LACS, he sees himself as an advocate rather than as an activist. Like so many animal protectionists I met during the course of my research, Bryant invokes the philosophy of Gandhi and nonviolence as the basic model for the animal rights movement: "To exact revenge and inflict physical and mental pain on those who abuse animals is to abandon the very principles upon which our cause is based. . . . If we stand for anything it is compassion, tolerance and reverence for all life" (1990:83). When I met Bryant at the league's open weekend, I was struck by his passion, moderation, and reasoned arguments, as well as by what appeared to be a leather coat! Bryant offers a combination of virtues, vices, and skills that is rare amongst true believers in the animal movement. He says that in common with many other animal activists, he joined the movement when *"someone handed me a leaflet."*

As a former hunt saboteur of some twenty years' experience, he has noticed how relations between the protagonists have changed for the worst since the 1960s, when there was *"a great deal of respect between the hunt saboteurs and hunters."* The "sabs" respected the skills of the hunters, who in turn acknowledged the hunt saboteurs as more credible than armchair critics. According to Bryant, it was during the skinhead and punk era in the 1980s that the relationship changed, and there were some nasty clashes between the protagonists. He sees the idea of hunting the hunted as counterproductive when it turns into violent confrontation, although he does pay tribute to the Hunt Saboteur Association (HSA) for its pioneering work in antihunting during the 1960s:

> *To be fair, when the league during the 1960s had virtually abandoned the political campaign and was just buying land up [for wildlife sanctuaries] . . . it was the hunt saboteurs that kept the issue of hunting in the press, and that is to their credit. But once people started to use violence, then it's been a disaster ever since for the whole movement really, because when there is a bombing, or anything like that, we find that our people are out in the streets with their tins collecting [and get told] "You're those bloody animal people!"* (Interview 1996)

The HSA is Britain's oldest direct-action animal protection SMO. It was set up in 1964 as a breakaway group from the league, which the hunt saboteurs believed had not achieved any significant legislative breakthroughs with its non-confrontational style. Hunt saboteurs began by using nonviolent confrontational tactics—harassing hunters by distracting hounds with whistles and the like—but the level of violence on both sides has increased in recent years. Hunters regard the hunt sab as "a curious amalgam of new age traveller, Class War anarchist and violent lefty" (Mann 1993:13). According to ACIG, at least three hunt saboteurs have died during protests in the 1990s, and one LACS member was murdered after an altercation with a man shooting birds. As a charity with over 40,000 members, LACS now dissociates itself from the HSA and says that being a "sab" is incompatible with the league's philosophy.

For Bryant and the league, violence is a tactical disaster, and even peaceful demonstrations are seen as counterproductive. He argues that London is today so congested with traffic that a peaceful march or demonstration only contributes to *"getting right up the noses of thousands of people who might be sympathetic to your cause but are going to go home very angry because they've been held up for an hour in the traffic."* If the demonstration turns violent, and *"somebody puts a brick through a window,"* then it becomes a tactical disaster, according to Bryant. He prefers achieving social change via leafleting, and he notes in his *Fettered Kingdoms* that the great strength of the animal rights movement lies with the supporters who hand out leaflets every week: "The leaflet is our media. In nearly twenty years in animal welfare and rights I have rarely found a campaigner who did not join the movement after being handed a leaflet—usually in the street" (Bryant 1990:88). He has nothing but praise for the foot soldiers who take their posts at the market stalls, and he is scathing in his condemnation of activists who break the law:

> *I wouldn't lift a finger to help anyone who's been jailed for violent animal rights activities. As far as I'm concerned they can throw the keys away. And I know Dave Calender, I've known him for years, and I know Ronny Lee, who went in for ten years, and I've met them many times. We've discussed the issues, and on the face of it, they are nice people. But it's easy to become violent in this world, and it's very hard to resist it. It's no good allowing yourself to be consumed by hatred. People are born into hunting, they're born into farming, they're born into all these animal exploitation things.* (Interview 1996) (Note: Calender and Lee were imprisoned for their animal rights activities.)

Bryant believes that the legislative route is ultimately the only effective means for achieving victory. Violence, on the other hand, is futile:

The trouble is, once people go down that route, then it's complete and utter failure. Because you're lined up against the whole of the state police, the court system, and if necessary, the whole of the armed forces. You're not going to win that battle because the one thing the state can do better than anyone else is violence. So not only is it totally immoral—because we're supposed to be a humanitarian cause, and in a democracy, we have a duty to use every militant but peaceful avenue up to the level of, and including the level of, civil disobedience—but any violence, intimidation, threats, abuse, particularly when it's targetted at individual researchers or individual huntsmen and people like that, then if we go down that way, there's no way back. (Interview 1996)

This is the philosophy that successfully combines the grassroots activities of activists equipped with cameras with the political skills of the league's organisational advocates who spend their time drafting animal welfare legislation. Thus, the league has no difficulty working with activists who engage in lawful, covert operations designed to break the back of the hunting fraternity. Huskisson's grassroots, anticruelty group's surveillance work with the league is a good example of effective cooperation between animal protectionists in the streets and in the suites. Huskisson's first undercover expose of animal cruelty was the Feldberg animal experimentation case, described in Melody McDonald's *Caught in the Act: The Feldberg Investigation* (1994) which was written in association with the ACIG.

MIKE HUSKISSON: THE CARING SLEUTH

Although the HSA and LACS share an undiluted opposition to hunting, they disagree about what are the best tactics to achieve a ban on blood sports. In recent years, LACS has used the services of the Animal Cruelty Investigation Group (ACIG) to expose the cruelty of the Hunt. The ACIG is the creature of a former hunt saboteur who, like Bryant, has since turned to more subtle and lawful tactics as a means of fighting the Hunt. Mike Huskisson founded the ACIG in 1989. From the outset, the undercover tactics have been effective in exposing cruelty in a number of contexts. Huskisson looks and sounds more like a gentlemanly scholar than someone who has spent time in prison for animal rights activism. He still sees himself as an activist working "in the field." He is the classic example of what Shapiro calls "the caring sleuth" (1994), an animal activist who actually seeks out instances of animal suffering and does something about it.

Huskisson first became aware of the Hunt while growing up in the countryside around Cambridge. What he saw convinced him that he had to do something to stop blood sports, and so he joined the HSA while a university student in London in 1971. He related an amusing story that illustrates how some journalists will even attempt to manufacture the news to reinforce the animal liberationist stereotype:

If they could find some person that would throw a brick at the front of the window, then they would highlight that. I always remember in about 1972 or 1973, I went to a demonstration against the parade of otter hounds of the Royal Windsor Horse Show, and the queen was there up in the royal box. And we were parading around the ring with banners protesting against the cruelty of otter hunting and this photographer and reporter from the Daily Express, I think it was, jumped into the ring and came in with me and he said to me, "Have you got any smoke canisters?" I said, "Oh, we've got one." He said, "Well, if you lob it up towards the royal box up there, I'll get a picture and put you on the front page." I said, "Well look, you give me the camera, you throw the smoke bomb, and I'll take your picture!" (Interview 1996)

An articulate, unassuming, and uninhibited interviewee, Huskisson's persuasive abilities are evidently feared by the hunting fraternity, who recently refused to debate him on the hunting issue at the Oxford Union on the pretext that his presence would taint their organisation (The British Field Sports Society), presumably a reference to his prison sentences several years ago.

The ACIG provides detailed tips to its one thousand seven hundred supporters throughout the country on how best to film transgressions of the Hunt. Grassroots activists are encouraged to take up the camera as the most effective weapon against people who abuse animals. According to Huskisson, anyone can use a camera to legally film the activities of the Hunt:

But now having had that experience [of illegal direct action], I say to young people, "No, the way to do it is to use your brains not your brawn, channel that anger into activity that's undercover and lawful. Get a video camera, get yourself a job in a research place, get yourself a job in a hunt kennel, go out there and get the film, and you're not breaking the law, but you're breaking the back of the opponents. (Interview 1996)

Huskisson's ACIG has been a pioneer of the tactic of undercover surveillance, in which activists investigate and seek to expose animal abusers. He maintains that this is more effective in *"breaking the back of opponents"* than violent and illegal activities:

And that's what we never had in the 1970s. There wasn't the direction from the senior echelons of the animal welfare/rights movement to say to young people, what you do with your anger. They were damn good at pumping up the anger with the images, and that led to people like myself and others ending up in prison because there wasn't any other outlet. . . . And to my mind, that's a crime. If you can get young lads between seventeen and twenty-five so angry that all they can do is go throw a brick through a window, then you've lost. It's a scandal, and what you've got to do is having got that anger generated, you have to show them how they can use it, how they can channel it, productively, in order to break the back of the opponents. (Interview 1996)

CRUELTY IN RECREATIONAL HUNTING

The ACIG's exposure of the Quorn Hunt, with which Prince Charles was associated, is one of the success stories of the LACS/ACIG teamwork and an illustration of the potency of effective activist/advocacy animal protectionism. Bryant acknowledges that there are worse forms of cruelty to animals than fox-hunting, such as vivisection and factory farming. Even so, he argues that if it is not possible to outlaw *"killing animals for fun,"* these other problems will never be resolved. To Bryant and Huskisson, banning hunting with hounds is a winnable issue, because as Huskisson points out in words reminiscent of Pacelle's, the public will not tolerate wanton cruelty:

People would react against overt cruelty when it's inflicted on an animal or a child. They will not allow it to happen. And the trouble is that the cruelty that goes on in blood sports and vivisection is hidden. But it is of the same kind. It's just like someone beating hell out of a dog with a stick. And we think that the public anger that would be whipped up by seeing it will be enough to put an end to it. (Interview 1996)

It is this moral capital or reservoir of goodwill in England toward animals that Huskisson hopes to capitalise on. The dislike of hunting is the reverse side of this goodwill, which for opponents of the Hunt takes the form of "hunting the hunters." I met several league members who said they actually loved harassing hunters with their cameras since there was little that the hunters could do to prevent ordinary people from recording what goes on. One supporter of the league described the video footage as "our main tool" against the Hunt:

There are many examples of us filming transgressions of the Hunt, showing the cruelty of the Hunt. They deny it is cruelty. We show the cruelty of the Hunt and it's resulted, for instance, in Prince Charles's hunt being banned, the Quorn Hunt. It has resulted in other hunts being banned. It's resulted in publicity over the national news and the national news showing our films. So we film the transgressions. We expose the lie. . . . Animals being seen on film to be torn apart. . . . We've seen photographs, for instance, of the stag being shot and injured, its jaw being blown off and the stag's still running, and at the end of the day, the stag having to be drowned in a river with a man putting his foot on its neck, pushing it under water. And this is illegal, this is against the Hunt rules. The Hunt, say for instance, in this area where we are talking now, where stag hunting is rife, they say it is a "clean kill." As I say, we have film of the stag's jaw being blown off and the stag getting up and running away. (Interview 1996)

For most of this century, the league has sought to convince legislators that just as the law recognises that domestic animals and animals in captivity are capable of suffering cruelty, so too should that principle be extended to wild animals. And for most of the century, the hunting fraternity has succeeded in thwarting their campaign since the implementation of the "cruelty" principle in law would effectively end blood sports in Britain. Much is at stake, therefore, in the prospect of a free parliamentary vote on the issue of hunting with hounds.

The British Field Sports Society (BFSS) has 85 000 members with 410 000 in affiliated clubs throughout the country. Impressive as these numbers are, they are more than matched by the combined membership of the animal welfare alliance of the RSPCA, the LACS, and IFAW, whose combined membership totals 590 000; an additional 925 000 supporters can be counted on, at least in theory, from the Royal Society for the Protection of Birds (RSPB).

These figures appeared in a feature article in *The Times* (10 August 1996) that asked their readers to reflect on who Britain's real animal lovers were—the hunting fraternity or the animal welfare lobby. Three representatives of the opposing sides were pictured giving their respective views on the issue: Carla Lane, a well-known English actress and "animal lover," Lord Mancroft, vice chairman of the BFSS, and James Barrington, ex-director of the LACS, who recently defected to the other side. The article acknowledges that there may have been some truth in the league's claim some weeks earlier that the hunting fraternity was on the run and was trying to divert attention away from hunting with hounds—the respectable side of the sport—to the more unpleasant activities of digging foxes out of their holes with terriers or attacking the animals with lurchers. Bryant claims in the league's journal that Barrington "is being used by top hunters as the instrument to save fox hunting by blaming the terrier and lurcher boys for the brutality of killing wild animals with dogs . . . [In other words] to sacrifice "working class" bloodsports so that traditional hunting with hounds can survive" (*Wildlife Guardian* Summer 1996:1).

As the threat of their demise seems inevitable at the turn of the century, the hunting fraternity has joined forces with a wider rural lobby, the Countryside Movement, in an endeavour to save hunting and other traditional rural pastimes. Rural folk have recently held large rallies in London to give city people their message that hunting is "our music, it is our poetry, it is our art, it is our pleasure . . . It is our whole way of life" (Baroness Anne Mallalieu's keynote address at the rally in 1997). Similar sentiments can be found on the BFSS Home Page on the Internet, in which hunters are described as "protectors" of animals, a concept that the antihunting lobby robustly rejects. On the other hand, hunters portray their opponents as "a cat and dog brigade" with little experience of animals apart from their pets.

Like the debates over animal experimentation, the quest for animal rights on the issue of hunting has been characterised by the rhetoric of vilification on both sides of the debate. Yet in their campaign against blood sports in the United Kingdom and in Australia as well, the animal protectionists do appear to be close to victory. As the enigmatic quote in the epigraph to this chapter suggests, the hedgehog may have the last word.

NOTE

1. First, he argues, most of the animals hunted in Australia are exotic or introduced species that are a threat to indigenous animals and even to the environment itself. Second, unlike their American and English counterparts, Australian hunters are not generally

perceived as pleasure seekers or sport hunters, but rather as workers who are keeping pests under control. Third, sympathy for hunters appears to be in inverse proportion to their social status, with poor people who hunt to supplement their diet being acceptable to the public while the recreational sport hunters are not. Franklin suggests that on these terms, Australian hunters score better than their English and American counterparts (1996:52–53).

PART IV

CONCLUSION: PROTECTING ANIMALS IN THE TWENTY-FIRST CENTURY

Nature teaches beasts to know their friends.

—William Shakespeare

Compassionate Beasts has focussed on animal activism and advocacy—the heart and mind of the movement, so to speak—in the seminal animal liberation campaigns against vivisection, factory farming, and sport hunting. Opposition to blood sports is the second-longest running campaign in the animal movement's history, and as predicted in the previous chapter, the one most likely to be won in the near future. Antivivisection has been the movement's oldest campaign and there is little doubt that it will remain the intractable, difficult issue that activists described throughout the book. Animal protectionists are divided on the issue of animal experimentation between hard-line abolitionists and pragmatists. In recent years, the issue has undergone a frame transformation in which the cruelty frame has been largely, though not entirely, supplanted by the claims of scientific antivivisectionism. From the perspective of this frame, antivivisectionists argue that animal experimentation is unscientific and a threat to human health. The spectre of genetic engineering of animals adds the threat of ecological catastrophe to this critique of animal experimentation as bad science.

A similar strategy of rationale expansion has been adopted by the opponents of commercial agriculture, which is also an arena for the improvement of food animals via genetic engineering. Animal activists claim that factory farming is not only inhumane, it also inflicts immense damage on the environment and poses great risks to human health. In expanding the movement's rationale in this way, the promotion of vegetarianism as an eco-friendly alternative to meat eating is seen by some animal protection organisations as a more positive and appealing message than the traditional anticruelty argument is. By appealing to human self-interest, animal rights/vegetarian advocates hope to mobilise greater numbers to support the cause of animal liberation.

The impact of genetic engineering on microorganisms, plants, animals, and humans offers the prospect of developments in these domains that few researchers could resist in the arenas of animal experimentation, commercial agriculture, and even hunting. Rather than being thought of as a new issue, genetic engineering is more likely to add a different, more sinister dimension to these three central issues in the animal movement's campaigns. Ironically, the genetic engineering of animals, as a technology that has the capacity to change the telos of an animal, may well increase the support for the animal movement among kindred spirits in ecology movements. The risk of ecological catastrophe that genetic engineering poses is so extreme that for the first time in recent history, it may be possible for people in disparate social movements and organisations to find common cause in the defence of nature (Munro 1999a).

I have suggested in Chapter 3 that while philosophers are important in the animal movement's moral crusade, its success ultimately depends on the actions of rank-and-file members and organisational leaders. Collective action is vital in all movements for social change; in a movement in which the beneficiaries are animals, organisation takes on even greater significance. The epigraph in Chapter 2 is pertinent here: *"If you want drama, get a movement; if you want results, get an organisation."* From the beginnings of collective action on behalf of animals, organisations like the RSPCA have built considerable reserves of moral capital and public goodwill by taking a moderate stance on the exploitation of animals. The opprobrium associated with cruelty to animals—and compassion as its ethical alternative—has resonated with the public in the three case study countries where the RSPCAs and SPCAs have been most active. As quintessential animal welfare organisations these protection societies have been successful in prosecuting individual cruelty cases, but not in undermining institutionalised cruelty to animals. Since the 1970s, some of the more dynamic animal rights/liberation organisations described in this book have taken up that task in the three case study countries.

COMPARING ANIMAL PROTECTION STRATEGIES IN THE UNITED STATES, UNITED KINGDOM, AND AUSTRALIA

In analysing the principal campaigns of these organisations, I have argued that English, American, and Australian animal protectionists use different combinations of advocacy/activist strategies in conducting their campaigns. These can be summarised as follows:

USA: Advocacy > Activism
UK: Activism > Advocacy
Australia: Activism + Advocacy

Animal protectionists in the United States are primarily interested in codifying public sentiment in the law via the constitutional route of organisational advocacy. By contrast, the grassroots mobilisation of public opinion and moral capital appears to be more characteristic of English animal protectionists. Their Australian counterparts tend to prefer a hybrid style of activism/advocacy in which the building of moral capital and animal welfare improvements in legislation go hand in hand. To be sure, animal protectionists in the three case study countries know that pity without entitlement is of little value. Whatever tactics are used, they must lead to improvements in animal welfare.

Many leading animal activists like Spira in America, Levy in Australia, and Court in the United Kingdom prefer to work alone or in small affinity groups. Their motivations are succinctly stated by Staggenborg when she notes that formalised organisational structures make it difficult for activists to take quick action or to bypass disputes over radical positions and confrontational tactics (1991:43). Activists like Spira, Levy, and Tyler in the United Kingdom have succeeded in their different campaigns because they have had the freedom to act swiftly and in ways, to use Tyler's words, that will *"shock and mesmerise and entice."*

In the United Kingdom, the tension between the amateurism of grassroots animal rights activism and the professionalism of organisational advocacy is much more in evidence than it is in the United States or in Australia. Even so, the relationship can be beneficial to both, as suggested in Fox's insightful observation in the epigram to Chapter 5: *"professionals keep the movement organized. Amateurs keep it honest."* Grassroots groups in the United Kingdom have traditionally felt frustrated by what can be achieved via constitutional routes to animal welfare. Exclusion from the issue networks and policy communities in the United Kingdom came to a head during the live export protests in 1995 and 1996 when activists wore T-shirts expressing their frustration thus: *"The Tory government won't listen so I have to shout!"* Consequently, in a nation of self-confessed animal lovers, grassroots activism has a much higher profile than in either the United States or Australia, where animal protectionists have greater access to policymakers.

I have argued in this book that the advocacy/activism relationship in the animal protection movement in Australia is, out of necessity rather than good management, more balanced than it is in the United Kingdom or the United States. I have also suggested that in Australia, more than in these two countries, animal activists historically have been able to forge alliances with kindred spirits in the conservation and environmental movements. These features are a function of the relatively small size of the animal movement in Australia, in terms of the organisations and the number of people who support them. As a consequence, the Australian animal movement lacks the resources available to its counterparts in the United States and the United Kingdom. Resources needed by social movement organisations include a large budget, a solid membership, computer equipment and databases, professional lobbyists, staff expertise on the relevant issues, a wide circle of contacts, a reputation for being credible and trustworthy, an appealing cause, and volunteer staff (Salazar 1996:633). Only the larger SMOs like the RSPCA or HSUS have all or most of these resources. With the exception of the last three resources that add up to reserves of moral capital, grassroots groups tend to have fewer material assets than their professional counterparts in the suites do. The hybrid activist/advocate groups in Australia do little better. ANZFAS is the only Australian organisation in this study that can claim to have access to most of these resources, although not a large budget and staff.

ANZFAS and its president, Singer, typify the unique combination of grassroots activism and organisational advocacy that characterises Australian animal protection. While it is itself primarily an animal advocacy, lobbying organisation, it serves as a federation of many other groups, most of which are grassroots, direct action in orientation. In this, ANZFAS is unique as an animal protection SMO. Thus, Animal Liberation branches in Australia, unlike their namesake counterparts elsewhere, have combined the tactics of grassroots activism—the spontaneous, uninhibited, exhibitionist stunts and moral rhetoric that come from the heart—with the strategy of institutional lobbying that organisational advocates like ANZFAS favour. Nonetheless, the movement as a whole in Australia could be improved by adapting some of the more successful features of animal protection SMOs elsewhere. For a start, Australia has many unpaid and enthusiastic amateurs in the movement but too few paid professionals.

PROFESSIONALISM IN ANIMAL PROTECTION

In the quest for greater professionalism, the AWI and AHA in the United States or CIWF in the United Kingdom might serve as a model for similar groups in Australia. This is not meant as a criticism of much older and experienced organisations such as WLPA, which was established a century ago. Like similar venerable animal protection organisations, WLPA's very survival is testimony to the strategy of "bearing witness." Simply by being there, as the antivivisectionists were during periods of movement decline, raises the movement's moral capital. Yet apart from the moral support of its members and the dedication of a tiny part-time staff, small groups like CADS, WLPA, and similar affinity groups elsewhere have little in the way of resources that professional advocacy organisations need to confront animal abusers.

In the Australian animal protection movement, there is not the same emphasis on fund-raising and organisational maintenance as there is in the United States and, to a lesser extent, in the United Kingdom. This is both good and bad for the movement in Australia. It is good in that the movement is unlikely to develop a "CEO culture," for which the larger environmental SMOs have been criticised. Success in fund-raising can become an end in itself and cause ostensibly altruistic organisations to forget their primary mission. On the other hand, a poorly resourced animal protection SMO that operates with voluntary part-time staff on a shoestring budget will be unable to make any impact on its more professionally organised adversaries. Virtually all of the Australian SMOs studied for this book fall into this category.

Although ANZFAS is professionally run, its budgetary limitations allow little room for expansion. For instance, ANZFAS's quarterly magazine, *Animals Today*, is of a high standard but lacks the in-depth pieces and coverage of issues—of say, *The Animals Agenda*—that only a full-time researcher/editor can provide. For smaller organisations for which in-house media are important, the position could perhaps be combined with the job of fund-raiser since virtually every SMO in the Australian sample had a small membership base. Critical mass, which movement leaders such as Regan see as crucial to the long-term success of the movement (Interview 1996), is lacking in Australia. It can best be built with the aid of paid professionals, rather than with the help of enthusiastic, amateur volunteers. Successful animal protection organisations use knowledge as an organisational weapon, along the lines deployed by the AWI in the United States, Animal Aid in the United Kingdom and ANZFAS in Australia. Enthusiastic amateurism is no longer an effective weapon against the more organised forces within the animal-user industries and the countermovements against animal liberation. Most of all, a more professional political lobby will be needed to counter the spectre of new developments in the genetic engineering of animals, which if unchecked, could unleash an ecological calamity of unprecedented proportions.

If the existence of countermovements can be interpreted as a sign of the initial social movement's success, then the quest for animal rights is on solid ground. I have argued elsewhere (Munro 1999b) that each of the animal movement's central campaigns has spawned virulent countermovements that seek to challenge the quest for animal rights. Countermovements in this backlash against animal liberation use some of the same tactics as their targets do, including the politics of emotion, the rhetoric of vilification, the claim of respectability, and so on. In at least one area, however, the animal movement is different to the countermovements, and that is in the gender balance. As I noted especially in Chapters 4 and 9, women constitute the vast majority of animal protectionists in welfarist, liberationist, and rightist organisations, as well as in veterinary science, although not in most other branches of science. Further research on the implications of these phenomena is needed, however, as only the most general explanations have been suggested in this book.

Finally, critics of the animal movement have attempted to demonise animal activists as violent extremists, a charge that I hope the testimony of the *compassionate beasts* represented in this book convincingly refutes. For all of the movement leaders in the three case study countries, the animal movement stands or falls on its ability to occupy what Singer calls "the high moral ground" (Singer, 1990:xiii). Violence is rejected by rank-and-file activists as well, including the proverbial "*brick through the window*" that came to symbolise the most publicised event in the quest for animal rights in recent history.

Bibliography

AAHR (1994). Australian Association for Humane Research Inc. 'Most Oppose Animal Experiments: Newspoll'. *Media Release*, 27 September.

AAHR (1997). Australian Association for Humane Research Inc. 'Fat-tailed Dunnart Experiments', Newsletter No. 75, November: 3.

ABC TV (1997). 'Animal Welfare', *7.30 Report*, 7 April.

Achbar, M. (1994). *Manufacturing Consent: Noam Chomsky and the Media*. Montreal: Black Rose Books,

Action (1993a). 'Paul Keating: Making Money from Pigs' Suffering', *Animal Liberation Magazine*, Issue 45: 8–9.

Action (1993b). 'Keating's Piggery To Quadruple in Size', *Animal Liberation Magazine*, Issue 46:17.

Action (1994). 'Parkville Piggery', *Animal Liberation Magazine*, Issue 50: 8–9.

Action (1995). 'Australia's Disgrace', *Animal Liberation Magazine*, Issue 53:10–11.

Action (1996). '300 Million Adolescents Murdered in the Breeding Sheds', *Animal Liberation Magazine* No. 56 (September).

Adam, B. (1993). Post-Marxism and the New Social Movements' in William K. Carroll (ed.) *Canadian Sociology and Anthropology Association* 30 (3) (August): 316–35.

Adams, C. (1990). *The Sexual Politics of Meat: A Feminist-Vegetarian Critical Theory*. Cambridge: Polity Press.

Adams, C., and Donovan, J. (1995). *Animals and Women: Feminist Theoretical Explorations*. Durham: Duke University Press.

Advertiser (1997). 'Tail End of Research into Weighty Problem', 26 March.

Alcorn, G. (1999). 'Assault and Battery', *The Age* 3 (February): 11.

Animal Aid (1996). 'Animal Aid: Past, Present and Future', Supplement to *Outrage*, February.

Animals' Agenda (1989a). 'A Need for Unity Now' (October): 2.

Animals' Agenda (1989b). 'Henry Spira' (July–August): 20–21.

Animals' Agenda (1996) 'Animal Rights Handbook' (July–August): 66–75.

Animals Today (1994–95). 'Animals and Ethics at Oxford', Interview with Andrew Linzey, 2 (4) (November–January): 14–16.

Animals Today (1995). 'Animal Liberation: The Book that Has Changed Perceptions about Animals', 3 (1) (February–April): 8–12.

Animals Today (1996) 'Intensive Piggeries: Confinement for Commerce', 4 (1) (February–April): 18–19.

ANZFAS (1994–95). Annual Report of the Australian and New Zealand Federation of Animal Societies, Melbourne.

Appleyard, B. (1995). 'A Moral Life in this Godless World', *The Independent* 13 (November): 19.

Arluke, A. (1990). 'Moral Elevation in Medical Research', *Advances in Medical Sociology* 1: 189–214.

———. (1991). Going into the Closet with Science: Information Control Among Animal Experimenters, *Journal of Contemporary Ethnography* 20:306.

Arluke, A., and Sanders, C. R. (1996). *Regarding Animals.* Philadelphia: Temple University Press.

Arluke, A., and Sax, B. (1992). 'Understanding Nazi Animal Protection and the Holocaust', *Anthrozoos* 5:6–31.

Australian Veterinary Association Ltd. (1993). 'Members' Directory and Policy Cromwell, J. R. and P. E. Greenwood (eds.) Compendium. Enfield, New South Water: Craft Printing Industries Pty. Ltd.

AWI (1996). *The Animal Welfare Institute Quarterly*, 45 (1) (Winter 1996).

Baker, S. (1993). *Picturing the Beast: Animals, Identity and Representation.* Manchester: Manchester University Press.

BALE (1995). Brightlingsea Against Live Exports, *Newsletter*, 10 October, Issue 4: 1–4.

Barker-Benfield, G. J. (1992). *The Culture of Sensibility: Sex and Society in Eighteenth Century Britain.* Chicago: The University of Chicago Press.

Barnard, N. D., and Kaufman, S. R. (1997). 'Animal Research is Wasteful and Misleading', *Scientific American* (February): 80–2.

Barnes, J. (1989). *A History of the World in 10½ Chapters.* London: Picador.

Barraclough, N., Neales, S., and Brew-Bevan, P. (1997). 'A Case of Neglect?', *Good Weekend, The Age* 13 December: 16–23.

Bekoff, M., and Meaney, C. A. (eds). (1998). *Encyclopedia of Animal Rights and Animal Welfare.* Westport, Connecticut: Greenwood Press.

Benton, T. (1992). 'Animal Rights and Wrongs: Prolegomena to a Debate', *Capitalism, Nature, Socialism: A Journal of Socialist Ecology* 3 (2) No. 10 (June): 79–82.

Benton, T., and Redfearn, R. (1996). 'The Politics of Animal Rights: Where is the Left?', *New Left Review* 215: 43–58.

Binding, P. (1995). 'Alive and Kicking', *New Statesman and Society* 13 (January):16–7.

Birke, L. (1994) *Feminism, Animals and Science: The Naming of the Shrew.* Buckingham: Open University Press.

———. (1995). 'Exploring the Boundaries: Feminism, Animals and Science' in C. Adams and J. Donovan (eds.), *Animals and Women: Feminist Theoretical Explorations*. Durham: Duke University Press.

———. (1997). 'Science and Animals or Why Cyril Won't Win the Nobel Prize', *Animal Issues* 1: 45–55.

Blackman, D. E., Humphreys, P. N., and Todd, P. (eds.) (1989). *Animal Welfare and the Law*. Cambridge: Cambridge University Press.

Blum, D. (1994). *The Monkey Wars*. New York: Oxford University Press.

Blue Mountains Gazette (1993). 'Regie Pushes for Eco-friendly Burger', 8 September.

Blue Mountains Whisper (1990). Regie McVegie for Mayor', October.

Botting, J. H., and Morrison, A. R. (1997). 'Animal Research is Vital to Medicine', *Scientific American* (February):83–5.

Braithwaite, J. (1989). *Crime, Shame and Reintegration*. Cambridge: Cambridge University Press.

Brunner, P. (1995). *Tierschutz in Deutschland und im Vereinigten Koenigsreich (Animal Protection in Germany and the United Kingdom*. PhD diss., University of Munich.

Bryant, J. (1990). *Fettered Kingdoms*. Manchester: Fox Press.

Budiansky, S. (1992). *The Covenant of the Wild: Why Animals Chose Domestication*. New York: William Morrow and Company, Inc.

Buechler, S. M., and Cylke, F. K. (1997). *Social Movements: Perspectives and Issues*. Mountain View, CA: Mayfield Publishing Company.

Buettinger, C. (1993). 'Antivivisection and the Charge of Zoophilia-Psychosis in the Early Twentieth Century', *The Historian* 55 (2): 277–88.

Canel, E. (1992). 'Social Movement Theory' in W. Carroll (ed.), *Organising Dissent: Contemporary Social Movements in Theory and Practice*. Toronto: Garamond Press.

Cartmill, M. (1993). *A View to a Death in the Morning: Hunting and Nature through History*. Cambridge, Massachusetts: Howard University Press.

Case, T. (1995). 'Editors, Cattlemen have a Beef', *Editor & Publisher* 128 (9) 4 (March): 9–11.

Catton, W. R., and Dunlap, R. E. (1980). 'A New Ecological Paradigm for Post-Exuberant Sociology', 24: 15–47.

Chase, M. P. (1990). 'Animal Rights: An Interdisciplinary Selection Bibliography', *Law Library Journal* 82: 359–91.

Clark, S. R. L. (1997). *Animals and Their Moral Standing*. London: Routledge.

Cockburn, A. (1996). 'A Short History of Meat', *New Left Review* 215: 16–42.

Coleman, V. (1993). 'Vivisection is Invalid', *Animals Today* (July–September): 16–7.

Collard, A., with Contrucci, J. (1988). *Rape of the Wild: Man's Violence Against Animals and the Earth*. Bloomington: Indiana University Press.

Cox, P. (1992). *The New Why You Don't Need Meat*. London: Bloomsbury Publishing Ltd.

Crook, S., Pakulski, J., and Waters, M. (1992). *Postmodernization: Change in Advanced Society*. London: Sage Publications.

Cusick, J. (1995). 'Sleepy Port that Joined Front Line in Trade War', *The Independent*, 20 January.

Davis, K. (1995). 'Thinking Like a Chicken: Farm Animals and the Feminist Connection' in C. Adams and J. Donovan (eds.), *Animals and Women: Feminist Theoretical Explorations*. Durham: Duke University Press.

Day, D. (1991). *The Eco Wars: A Layman's Guide to the Ecology Movement*. London: Paladin.

Deal, C. (1993). *The Greenpeace Guide to Anti-Environmental Organisations*. Berkeley: Odonian Press.

de Fraga, C. (1993). 'Cruelty in the Name of Culture', *Animal Liberation Magazine* 13 January-March: 16–17.

Dizard, J. E. (1994). *Going Wild: Hunting, Animal Rights and the Contested Meaning of Nature*. Amherst: University of Massachusetts Press.

Donald, R. (1998). 'Welcome to WSPA', *Animals Today* 6 (3) (August/October): 1–13.

Donovan, J. (1993). 'Animal Rights and Feminist Theory' in G. Gaard (ed.), *Ecofeminism: Woman, Animals, Nature*. Philadelphia: Temple University Press.

Dowie, M. (1995). *Losing Ground: American Environmentalism at the Close of the Twentieth Century*. Cambridge, MA: The Massachusetts Institute of Technology Press.

Dunayer, J. (1995). 'Sexist Words, Sexist Roots' in C. Adams and J. Donovan (eds.), *Animals and Women: Feminist Theoretical Perspectives*. Durham: Duke University Press.

Eagar, C. (1995). 'The Middle Class Goes Militant', *The Observer*, 22 January: 5.

Economist (1995a). 'Also a Part of Creation', 19–25 August: 17–19.

Economist (1995b). 'The Meat of the Matter', 21 January: 62.

Economist (1995c). 'What Humans Owe to Animals', 19–25 August: 11–12, 17–19.

Eckersley, R. (1992). *Environmentalism and Political Theory: Toward an Ecocentric Approach*. London: UCL Press.

Eder, K. (1990). 'The Cultural Code of Modernity and the Problem of Nature: A Critique of the Naturalistic Notion of Progress' in J. Alexander and P. Sztompka (eds.), *Rethinking Progress: Movements, Forces and Ideas at the End of the Twentieth Century*. London: Unwin Hyman.

Elias, N. (1978). *The Civilising Process, Vol 1, The History of Manners*. Oxford: Basil Blackwell.

Elliott, A. (1992). 'People Working for Animals', *Animal Liberation Magazine* No. 42 (October-December): 14–16.

Elston, M. (1990). 'Women and Anti-Vivisection in Victorian England, 1870–1900' in N. Rupke (ed), *Vivisection in Historical Perspective*. London: Routledge.

———. (1992). 'Caught in the Crossfire of the Animal Wars', *New Scientist* 23 (May): 32–5.

Erlichman, J. (1995). "Rent-a-mob" Forces End to Animal Welfare Protests', *The Guardian*, 5 January: 6.

Essex County Standard (1995). 'How a Community Hit the Headlines When a Port Was Used to Export Live Animals', 20 January.

Evans, E. P. (1906). *The Criminal Prosecution and Capital Punishment of Animals: The Lost History of Europe's Animal Trials*. London: William Heinemann.

Eyerman, R., and Jamison, A. (1989). 'Environmental Knowledge as an Organizational Weapon', *Social Science Information* 28 (1): 99–119.

FARM (1996). Farm Animal Reform Movement, *The FARM Report*, World Animal Awareness Week, 1996.

Ferry, L. (1995). *The New Ecological Order*. Chicago: The University of Chicago Press.

Fiddes, M. (1991). *Meat: A Natural Symbol*. London: Routledge

Finsen, L., and Finsen, S. (1994). *The Animal Rights Movement in America: From Compassion to Respect.* New York: Twayne Publishers.

Flacks, R. (1988). *Making History: The American Left and the American Mind.* New York: Columbia University Press.

Fleming, D. ([1954] 1987). *William H. Welch and the Rise of Modern Medicine.* Baltimore: John Hopkins University Press.

Fowler, F. (1988). *Survey Research Methods.* Newbury Park: Sage Publications.

Francione, G. (1996). *Rain Without Thunder: The Ideology of the Animal Rights Movement.* Philadelphia: Temple University Press.

Franklin, A. (1996). 'Australian Hunting and Angling Sports and the Changing Nature of Human-Animal Relations in Australia', *Australian and New Zealand Journal of Sociology* 32 (3) (November): 39–56.

French, R. D. (1975). *Antivivisection and Medical Science in Victorian Society.* Princeton: Princeton University Press.

Frey, R. G. (1980). *Interests and Rights: The Case Against Animals.* Oxford: Clarendon Press.

———. (1983). *Rights, Killing and Suffering: Moral Vegetarianism and Applied Ethics.* Oxford: Basil Blackwell.

Gamson, W. (1985). 'Goffman's Legacy to Political Sociology', *Theory and Society* 14: 551–604.

———. (1992a). *Talking Politics.* Cambridge: Cambridge University.

———. (1992b) 'The Social Psychology of Collective Action' in A. D. Morris, and C. M. Mueller (eds.), *Frontiers in Social Movement Theory.* New Haven: Yale University Press.

Garner, R. (1993). *Animals, Politics and Morality.* Manchester: Manchester University Press.

———. (1995). 'The Politics of Animal Protection', *Society of Animals* 3 (1): 43–60.

George, K. P. (1994). 'Should Feminists be Vegetarians?', *Signs* Winter: 405–34.

Gilligan, C. (1992). *In a Different Voice: Psychological Theory and Women's Development.* Cambridge: Harvard University Press.

Goffman, E. (1974). *Frame Analysis.* New York: Harper and Row.

Gordon, L. (1994). *Pitied but Not Entitled: Single Mothers and the History of Welfare.* Cambridge: Harvard University Press.

Gregory, M. (1994). 'Commerce and Cruelty', *Alternative Law Journal* 19 (4) (August): 86–9.

Groves, J. M. (1995). 'Learning to Feel: The Neglected Sociology of Social Movements', *The Sociological Review* 43 (3): 435–61.

———. (1997). *Hearts and Minds: The Controversy over Laboratory Animals.* Philadelphia: Temple University Press.

Guardians Newsletter (1994). 'Peter Singer: Lab Animal's Friend or Foe?', November.

Guardians Newsletter (1995). 'Animal Rights: The Suicide of a Movement', August.

Guardians Newsletter (1996). 'Animal Rights, Animal Welfare: The Hijacking of a Movement', Summer.

Guither, H. (1998). *Animal Rights: History and Scope of a Radical Social Movement.* Carbondale: Southern Illinois University Press.

Gunn, C., and Gunn, H. (1991). *Reclaiming Capital: Democratic Initiatives and Community Development.* New York: Cornell University Press.

Haltunen, K. (1995). 'Humanitarianism and the Philosophy of Pain in Anglo-American Culture', *The American Historical Review* 100 (2): 303–34.

Harraway, D. (1989). *Primate Visions: Gender, Race and Nature in the World of Modern Science*. New York: Routledge.

Harries-Jones, P. (ed) (1991). *Making Knowledge Count: Advocacy and Social Science*. Montreal: McGill-Queens University Press.

Harris, M. (1985). 'The Animal Rights Brigade', *New Society* 31 (January): 168–71.

Harrison, B. (1973). 'Animals and the State in Nineteenth Century England' in J. M. Wallace-Hadrill and J. M. Roberts (eds.), *The English Historical Review* vol. 88 (349): 786–820.

Hellerstein, A. (1994). 'Activists Target Animal Dealers', *The Physiologist* 37 (6): 281–4.

Hendrick, G., and Hendrick, W. (eds) (1989). *The Savour of Salt: A Henry Salt Anthology*. Sussex, Fontwell: The Centaur Press.

Herman, E., and Chomsky, N. (1988). *Manufacturing Consent: The Political Economy of the Mass Media*. New York: Pantheon.

Herzog, H. (1993). 'Thoughts on the Nazi Animal Protection Movement', *Anthrozoos* 6 (2): 82–4.

———. (1995). 'Has Public Interest in Animal Rights Peaked?', *American Psychologist* 50 (11) (November): 945–7.

Hill, L. (1996). *The Case for Vegetarianism: Philosophy for a Small Planet*. Lanham, MD: Rowan and Littlefield Publishers.

Hills, C. (1995). 'Essex Police Back Up Fight for Order in Animal Demo', *East Anglian Daily News*, 9 January.

Hitt, J. (1988). 'Just Like Us? Toward a Notion of Animal Rights', *Harper's Magazine* (August): 44–52.

Hogan, W., and Retzel, E. (1995). 'Ethical Treatment of Animals', *Choice* (September): 59–72.

Jamieson, D. (1985). 'Against Zoos' in P. Singer (ed.), *In Defense of Animals*. New York: Harper and Row Publishers.

Jasper, J., and Nelkin, D. (1992). *The Animal Rights Crusade: The Growth of a Moral Protest*. New York: The Free Press.

Jasper, J., and Poulsen, J. (1995). 'Recruiting Strangers and Friends: Moral Shocks and Social Networks in Animal Rights and Anti-Nuclear Protests', *Social Problems* 42 (4): 493–512.

Jennett, C., and Stewart, R. (1989). *Politics of the Future: The Role of Social Movements*. Melbourne: Macmillan.

Johnson, L. (1993). 'Review of M. Leahy's Against Liberation', *Anzccart News* 6 (4).

Guardians Newsletter (1994). 'Peter Singer: Lab Animal's Friend or Foe?', November.

Guardians Newsletter (1995). 'Animal Rights: The Suicide of a Movement', August.

Guardians Newsletter (1996). 'Animal Rights, Animal Welfare: The Hijacking of a Movement', Summer.

Kant, I. (1930). *Lectures on Ethics*. London: Methuen.

Katz, J. (1988). *Seductions of Crime: Moral and Sensual Attractions of Doing Evil*. New York: Basic Books.

Kellert, S. R. and Westervelt, M. D. (1982). 'Historical Trends in American Animal Use and Perception', *Transactions of the Forty-seventh North American Wildlife and Natural Resources conference* 47: 649–663.

Kennedy, J. S. (1992). *The New Anthropomorphism.* Cambridge: Cambridge University Press.

Kerasote, T. (1993). *Bloodties: Nature, Culture and the Hunt.* New York: Random House.

Komesaroff, P. (1992). 'Bioethics and Nature: The Case of Animal Experimentation', *Thesis Eleven* 32: 65–75.

Kuhse, H. (1992). 'Animals That are Less Equal Than Some', *The Australian* 27 (November): 13.

Kunkel, K. (1995). 'Down on the Farm: Rationale Expansion in the Construction of Factory Farming as a Social Problem' in J. Best (ed.), *Images of Issues: Typifying Contemporary Social Problems.* New York: Aldine De Gruyter.

Kysar, V. (1996). 'The Shameful Hunt', *Libearty* Summer 53: 22–3.

La Follette, H., and Shanks, N. (1996). *Brute Science: Dilemmas of Animal Experimentation.* London: Routledge.

Leahy, M. (1991). *Against Liberation: Putting Animals in Perspective.* London: Routledge.

Leahy, M., and Cohn-Sherbok, D. (1996). *The Liberation Debate: Rights at Issue.* London: Routledge.

Lederer, S. (1990). 'The Controversy over Animal Experimentation in America, 1880–1914' in N. Rupke (ed.), *Vivisection in Historical Perspective.* London: Routledge.

———. (1991). 'Dogfights: The Use of Pet Animals in Biomedical Research 1938–1966', Paper presented at the meeting of the American Association for the History of Medicine, Cleveland, Ohio, May 1991.

———. (1992). 'Political Animals. The Shaping of Biomedical Research Literature in Twentieth-Century America', *ISIS* 83: 61–79.

Lemert, J. B. (1984). 'News Context and the Elimination of Mobilizing Information: An Experiment', *Journalism Quarterly* Summer: 243–49.

Lewis, C. S. ([1947] 1971). 'Vivisection' in W. Hooper (ed.), *Undeceptions: Essays on Theology and Ethics.* London: Geoffrey Blés.

Lichterman, P. (1996). *The Search for Political Community: American Activists Reinventing Commitment.* Cambridge: Press Syndicate of the University of Cambridge.

Lines, W. J. (1992). *Taming the Great South Land: A History of the Conquest of Nature in Australia.* Sydney: Allen and Unwin.

Linzey, A. (1994). *Animal Theology.* Urbana: University of Illinois Press.

Lyons, Z. D. (1996). 'Editorial Reply', *Boycott Quarterly* 3 (4) Summer: 4–7.

McAdam, D. (1986). 'Recruitment to High Risk Activism: The Case of Freedom Summer', *American Journal of Sociology.* 92: 64–90.

———. (1988). 'Micromobilization Contexts and Recruitment to Activism', *International Social Movement Research* 1:125–154.

McAdam, D., McCarthy, J., and Zald, M. (1988). "Social Movements" in N. Smelser (ed), *Handbook of Sociology.* Beverley Hills: Sage.

McAdam, D., and Paulsen, R. (1993). 'Specifying the Relationships Between Social Ties and Activism', *American Journal of Sociology* 99 (3) (November): 640–67.

McCarthy, J., and Zald, M. N. (1973). *The Trend of Social Movements in America: Professionalization and Resource Mobilization.* Morristown, NJ: General Learning Press.

MacCulloch, J. (1993). *Creatures of Culture: The Animal Protection and Preservation Movements in Sydney, 1880–1930*. PhD thesis. University of Sydney.

McDonald, M. (1994). *Caught in the Act: The Feldberg Investigation*. Oxford: Jon Carpenter Publishing.

McGuinness, P. (1993a). 'World Whaling Ban Survives on Myths Alone', *The Australian*, 14 April.

———. (1993b). 'Lifting Whaling Ban a Test of Sentiment Versus Sanity', *The Australian*, 17–18 April.

Maehle, A. H. and Troehler, U. (1990). 'Animal Experimentation from Antiquity to the End of the Eighteenth Century: Attitudes and Arguments' in A. R. Nicholaas (ed.), *Vivisection in Victorian Perspective*. London: Routledge.

Magel, C. (1989). *Keyguide to Information Sources in Animal Rights*. London: Mansell Publishing.

Malcolm, B., Sale, P., and Egan, A. (1996). *Agriculture in Australia: An Introduction*. Melbourne: Oxford University Press.

Mann, N. (1993). 'Not Giving a Fox for the Sabbers,' *New Statesman & Society*, 19 November: 12–13.

Margalit, A. (1996). *The Decent Society*. Cambridge, Massachusetts: Harvard University Press.

Marks, K. (1995) 'Q: What Made the Tory Businesswoman Hurl Abuse at Lorry Drivers for 5 Nights? A: The Sight of Live Calves Heading for the Veal Crates', *The Weekly Telegraph*, 11–17 January: 11.

Marsh, I. (1994). 'Interest Groups Analysis' in A. Parkin, J. Summers, and D. Woodward (eds.), *Government, Politics and Policy in Australia*. 5th ed. Melbourne: Longman Cheshire.

Masson, J., and McCarthy, S. (1995). *When Elephants Weep: The Emotional Lives of Animals*. New York: Delacorte Press.

Mathews, L., Loveridge, A. M., and Guerin, B. (1994). *Animal Welfare Issues and Attitudes in New Zealand*. Hamilton, New Zealand: Animal Behaviour and Welfare Research Centre.

Midgley, M. (1983). *Animals and Why They Matter*. Harmondsworth: Penguin.

Moore, C. (1995). 'Now the Protesters Have a New Group of Creatures Who Will Never Answer Back', *The Spectator*, 14 January: 7.

Munro, L. (1993/1994). 'A Decade of Animal Liberation', *Current Affairs Bulletin* 70 (7): 12–19.

Munro, L. (1994). 'Grassroots Animal Rights Activism: The Case of Mountain Residents for Animal Rights', *Journal of Regional Issues* 28: 37–48.

Munro, L.(1996a). *Animals and Social Issues Survey*. Questionnaire 1–8 available from lyle.munro@arts.monash.edu.au.

———. (1996b). *Mapping the Animal Movement: A Survey of Animal Rights and Welfare Advocates in Australia*. Paper presented at the International Society of Anthrozoology Conference, Cambridge, Downing College, 24–26 July.

———. (1997a). 'Framing Cruelty: The Construction of Duck Shooting as a Social Problem', *Society & Animals* 5 (2): 137–54.

———. (1997b). 'Narratives of Protest: Television's Representation of an Animal Liberation Campaign', *Media International Australia* 83: 103–12.

———. (1999a). 'From Vilification to Accommodation: Making a Common Cause Movement', *Cambridge Quarterly of Healthcare Ethics* 8:46–57.

———. (1999b). 'Contesting Moral Capital in Campaigns Against Animal Liberation', *Society & Animals* 7: 35–53.

Neale, G. (1995). 'Why the Elderly Are Revolting', *The Weekly Telegraph*, Issue No. 197:21.

New Statesman & Society (1995a). 'England Rising: Have We Entered a New Age of Protest?' 27 January:5.

New Statesman & Society (1995b). 'Ban the Trade, Not the Protests', Editorial, 21 April: 5.

Nicoll, C. Russell, S. and Lau, A. (1992). 'Animal Liberation: An Exchange', *The New York Review*, 5 November:59.

Nilsen, A. P. (1977). 'Sexism as Shown Through the English Vocabulary' in A. P. Nilsen, H. Bosmajian, H. Gershuny, and J. Stanley (eds.), *Sexism and Language*. Urbana, IL: National Council of Leaders of English.

Nilsson, G. (1980). *Facts about Furs*. Rev. ed. Washington DC: Animal Welfare Institute.

Noske, B. (1989). *Humans and Other Animals*. London: Pluto Press.

Oliver, P., and Marwell, G. (1992). 'Mobilising Technologies for Collective Action' in A. D. Morris and C. McClurg-Mueller (eds.), *Frontiers of Social Movement Theory*. New Haven: Yale University Press.

Oogjes, G. (1991). 'Equine Welfare in Competitive Events Other Than Racing', Report of the Senate Select Committee on Animal Welfare, August, Commonwealth of Australia.

———. (1997). 'To Regulate or Not To Regulate?: The View of the Animal Rights Movement', RSPCA Australian Scientific Seminar, 25 February: 1–11.

———. (1997). 'Codes: A Painful Mistake?', *Animals Today* 5 (2) (May-July): 23–25.

Oogjes, G., and Page, N. (1996). 'Directions', *Animals Today* 4 (3) (August-October): 5.

Orlans, F. B. (1993). *In the Name of Science: Issues in Responsible Animal Experimentation*. Oxford: Oxford University Press.

Ortega y Gasset, J. (1972). *Meditations on Hunting*. trans. H. B. Wescot. New York: Scribner's.

Outrage (1996). *The Animal Journal for Action* 104 (October–November): 2.

Pacelle, W. (1996). 'Another State Ballot Victory: Californians Reject Trophy Hunting of Lions', *HSUS News* 41 (3), Summer.

Pakulski, J. (1991). *Social Movements: The Politics of Moral Protest*. Melbourne: Longman Cheshire.

Palmer, A. (1995). 'Carnival of Animal Libbers', *The Spectator*, 21 January: 16.

Pearson, M. (1996). 'Intensive Piggeries: Confinement for Commerce', *Animals Today* (February-April): 18–9.

———. (1997). 'Bunge Piggery—City of Despair', *Animal Action* Summer: 6–7.

Penman, D. (1995). '1000 Police Drafted to Animal Port', *The Independent*, 5 January.

Phillips, M. T., and Sechzer, J. A. (1989). *Animal Research and Ethical Conflict: An Analysis of the Scientific Literature, 1966–1986*. New York: Springer-Verlag.

Plous, S. (1996). 'Attitudes Toward the Uses of Animals in Psychological Research and Education: Results from a National Survey of Psychologists', *American Psychologist* 51 (11): 1167–80.

Plumwood, V. (1997). 'Babe: The Tale of the Speaking Meat', *Animal Issues: Philosophical and Ethical Issues related to Human/Animal Interactions* 1 (1): 21–36.

Primatt, H. ([1776] 1992). *The Duty of Mercy and the Sin of Cruelty to Brute Animals.* Sussex, Fontwell: Centaur Press. Reprint edited by R. Ryder.

Project Jonah (1977). 'Save the Whale!', *Wildlife in Australia* 14 (2) (June): 44–9.

Regan, T. (1984). *The Case for Animal Rights.* London: Routledge.

———. (1987). *The Struggle for Animal Rights.* Clarks Summit, PA: International Society for Animal Rights Inc.

Richards, R. ([1990] 1992). *Consensus Mobilization through Ideology, Networks and Grievances: A Study of the Contemporary Animal Rights Movement.* PhD diss.

Richards, S. (1990). 'Vicarious Suffering, Necessary Pain: Physiological Method in Late Nineteenth Century Britain' in N. A. Rupke (ed.), *Vivisection in Historical Perspective.* London: Routledge.

Ridley, M. (1995). 'Bunny-huggers Do Battle', *The Weekly Telegraph,* 18–24 January, Issue No. 183.

Rissel, C. and Douglas, W. (1993). 'Environmental Issues as Prime Time Television', *Media Information Australia,* 68, May: 86–92.

Ritvo, H. (1987). *The Animal Estate: The English and Other Creatures in the Victorian Age.* Cambridge, Massachusetts: Harvard University Press.

Roach Anleu, S. (1993). *Deviance, Conformity and Control.* Melbourne: Longman Cheshire.

Robbins, J. (1987). *Diet for a New America.* Walpole, NH: Stillpoint Publishing.

———. (1995). 'The Great Divide', *E Magazine* 7, (1) (February): 36–41.

Rochon, T. R. (1990). 'The West European Peace Movement and the Theory of New Social Movements' in R. Dalton and M. Kuechler (eds.), *Challenging the Political Order: New Social and Political Movements in Western Democracies.* Cambridge: Polity.

Rodd, R. (1992). *Biology, Ethics and Animals.* Oxford: Oxford University Press.

Roderick, S. (1997). 'No Mountain Too High', *The Protector* Winter: 10–11.

Rogers, L., and Tremont, R. (1992). 'Australian Researchers Take on Animal Rights', *New Scientist* 16 (May): 37.

Rootes, C. (1984). 'Protest, Social Movements, Revolution', *Social Alternatives* 4 (1): 4–8.

Rowan, A. (1989). 'Ethical Dilemmas in Experimentation' in D. Paterson and M. Palmer (eds.), *The Status of Animals: Ethics, Education and Welfare.* Wallingford: CAB International.

———. (1997). 'Forum: The Benefits and Ethics of Animal Research', *Scientific American* (February): 79.

Rowan, A., and Shapiro, K. (1996). 'Animal Rights, a Bitter Apple', *American Psychologist* 51 (11) (November): 1183–4.

Rucht, D. (1996). 'The Impact of National Contexts on Social Movement Structures: A Cross-Movement and Cross-National Comparison' in D. McAdam, J. McCarthy, and M. Zald (eds.), *Comparative Perspectives on Social Movements: Political Opportunities, Structures and Cultural Framings.* New York: Cambridge University Press.

Ruddick, S. (1980). 'Maternal Thinking', *Feminist Studies,* 6: 350–51

Ruedig, W. (1995). 'Between Moderation and Marginalization: Environmental Radicalism in Britain' in B. Taylor (ed.), *Ecological Resistance Movements: The Global Emergence of Radical and Popular Environmentalism.* New York: The State University of New York.

Rupke, N. (ed.) (1990). *Vivisection in Victorian Perspective*. London: Routledge.
Ryder, R. (1979). 'The Struggle Against Speciesism' in D. Paterson and R. Ryder (eds.), *Animal Rights—A Symposium*. Sussex, Fontwell: Centaur Press Ltd.
———. (1983). *Victims of Science: The Use of Animals in Research*. London: National Anti-Vivisection Society Ltd.
———. (1989). *Animal Revolution: Changing Attitudes Towards Speciesism*. Oxford: Basil Blackwell.
Salazar, D. J. (1996). 'The Mainstream Grassroots Divide in the Environmental Movement: Environmental Groups in Washington State', *Social Science Quarterly* 77 (3) (September): 626–43.
Salt, H. ([1892]) 1980. *Animals' Rights Considered in Relation to Social Progress: With a Bibliographical Appendix*. New York: Macmillan.
Sanderson, J. B. (ed.) (1873). *Handbook for the Physiological Laboratory*. London: Churchill.
SAPL, The (1997). 'The Work of the Society for Animal Protective Legislation', The SAPL Home Page, *http://www.animalwelfare.com/sapl*: 1–5.
Scarce, R. (1990). *Eco-Warriors: Understanding the Radical Environmental Movement*. Chicago: Noble Press.
Schiebinger, L. (1993). *Nature's Body: Sexual Politics and the Making of Modern Science*. London: Pandora.
Schlosberg, D. (1995). 'Communicative Action in Practice: Intersubjectivity and New Social Movements', *Political Studies* 43 (2): 291–311.
Schudson, M. (1989). 'How Culture Works: Perspectives from Media Studies on the Efficacy of Symbols', *Theory and Society* 18: 153–80.
Scruton, R. (1996). *Animal Rights and Wrongs*. London: Demos.
Senate Select Committee Report on Animal Welfare (1985). *Export of Live Sheep from Australia*. Canberra: AGPS.
———. (1988). *Kangaroos*. Canberra: AGPS.
———. (1989). *Animal Experimentation*. Canberra: AGPS.
———. (1990). *Intensive Livestock Production*. Canberra: AGPS.
Serpell, J. (1986). *In the Company of Animals*. Oxford: Basil Blackwell.
Shaiko, R. G. (1993). 'Greenpeace USA: Something Old, New, Borrowed', *The Annals of the American Academy of Political and Social Science* 528 (July): 88–100.
Shapiro, K. (1994). 'The Caring Sleuth: Portrait of an Animal Rights Activist', *Society and Animals* 2 (2): 145–65.
Shaw, R. (1996). *The Activist's Handbook: A Primer for the 1990s and Beyond*. Berkeley: University of California Press.
Sheppard, V. (1979). *My Head Against the Wall: A Decade in the Fight Against Blood Sports*. Wiltshire: Moonraker Press.
Silverstein, H. (1996). *Unleashing Rights: Law, Meaning and the Animal Rights Movement*. Ann Arbor: University of Michigan.
Singer, P. (1975). *Animal Liberation: A New Ethics for our Treatment of Animals*. New York: Random House.
———. (1985). 'Ten Years of Animal Liberation', *New York Review*: 46–52.
———. (1990). *Animal Liberation* 2d ed. London: Jonathon Cape.
———. (1992). 'Animal Liberation: An Exchange', *The New York Review* 5 November: 59.
———. (1995a). 'Animal Liberation: The Book that Changed Perceptions about Ani-

mals—an Interview with the Author', *Animals Today* 3 (4) (February–April): 8–12.

———. (1995b). 'Guns and Animals: An Animal Liberation Perspective' in J. Crook (ed)., *Hunting: A Critical Perspective*. Melbourne: Gun Control Australia, Inc.

———. (1997). *Henry: One Man's Way*, a film written and produced by Peter Singer and co-produced with John Swindells.

———. (1998). *Ethics into Action: Henry Spira and the Animal Rights Movement*. Lanham, MD: Rowan and Littlefield Publishers Ltd.

Smith, M. J. (1993). *Pressure, Power and Policy: State Autonomy and Policy Networks in Britain and the United States*. New York: Harvester Wheatsheaf.

Snow, D. A., Burke Rochford, E. Jr., Worden, S. K., and Benford, R. D. (1986). 'Frame Alignment Processes, Micromobilization and Movement Participation', *American Sociological Review* 51: 464–81.

Snow, D. A., and Benford, R. D. (1988). 'Ideology, Frame Resonance and Participant Mobilization' in B. Klandermans, H. Kriesi, and S. Tarrow (eds.), *From Structure to Action: Comparing Social Movement Research Across Cultures*. Greenwich, CT: JAI Press.

Spencer, C. (1993). *The Heretic's Feast: A History of Vegetarianism*. London: Fourth Estate.

Staggenborg, S. (1991) *The Pro-Choice Movement: Organization and Activism in the Abortion Conflict*. New York and Oxford: Oxford University Press.

Staggenborg, S. (1988). 'The Consequences of Professionalization and Formalization in the Pro-Choice Movement', *American Sociological Review* 53 (August): 585–606.

Stevens, C. (1996a). 'Historical Motivation for the Federal Animal Welfare Act: An Address on the Occasion of the 30th Anniversary of the Animal Welfare Information Center', unpublished paper, USDA Center, Riverdale, Maryland, 12 September: 1–18.

———. (1996b). 'The Leghold Trap', *Animal Welfare Institute Quarterly* 45 (1).

Sutherland, A., and Nash, J. E. (1994). 'Animal Rights as a New Environmental Cosmology' in *Qualitative Sociology* 17 (2): 171–86.

Swan, J. A. (1995). *In Defense of Hunting*. San Francisco: Harper.

Sznaider, N. (1996). 'Pain and Cruelty in Socio-Historical Perspective', *International Journal of Politics, Culture and Society* 10 (2): 331–54.

Talbot, J. (1981). *Whales*. Melbourne: Greenhouse Publications Pty. Ltd.

Tanenbaum, E. (1995). *Animal Transport Through Brightlingsea: Report of an Opinion Survey*. Essex: Department of Government, University of Essex.

Tannenbaum, J. (1993). 'Veterinary Medical Ethics: A Focus on Conflicting Interests', *Journal of Social Issues* 49: 143–56.

Tarrow, S. (1994). *Power in Movement: Social Movements, Collective Action and Politics*. Cambridge: Cambridge University.

Taylor, T. (1772) *A Vindication of the Rights of Beasts*. London: Eduard Jefferey.

Taylor, V. (1989). 'Social Movement Continuity: The Women's Movement in Abeyance', *American Sociological Review* 54 (5): 761–75.

Tester, K. (1989). *The Relationship Between the Idea and the Practice of Animal Rights*. PhD diss. University of Leeds.

———. (1991). *Animals and Society: The Humanity of Animal Rights*. London: Routledge.

Thomas, K. (1983). *Man and the Natural World: Changing Attitudes in England 1500–1800*. Middlesex, England: Penguin Books.

Thompson, F. M. L. (1988). *The Rise of Respectable Society: A Social History of Victorian Britain 1830–1900*. Cambridge, Massachusetts: Harvard University Press.

Thompson, T. (1996). 'Fur is Flying at the Humane Society', *Washington Post*, 14 August: A14.

Times, The. (1996). 'Who are Britain's real animal lovers?' 10 August: 21–22.

Townend, C. (1981). *A Voice for the Animals: How Animal Liberation Grew*. Sydney: Kangaroo Press.

———. (1993). 'India: Religion, Culture and Animals', *Animal Liberation Magazine* (April-June): 12.

———. (1995). 'Working for Animals in India', *Animals Today* (February-April): 24–5.

Turner, E. S. ([1964] 1992). *All Heaven in a Rage*. Fontwell, Sussex: Centaur Press.

Turner, J. (1980). *Reckoning with the Beast: Animals, Pain and Humanity in the Victorian Mind*. Baltimore: The Johns Hopkins University Press.

Turner, R., and Killian, L. (1987). *Collective Behavior*. 3rd ed. Englewood Cliffs, New Jersey: Prentice-Hall Inc.

Tyler, A. (1995). 'Silence of the Lambs', *An Animal Aid Investigation*. Tonbridge: Animal Aid.

———. (1996). 'No Defence: Support Grows for Our Campaign', *Outrage* 104 (October-November): 12–3.

Vidal, J. (1997) *McLibel: Burger Culture on Trial*. London: Macmillan.

VSUK (Vegetarian Society of the United Kingdom) (1996). 'Laying Hens: Information Sheet', Internet home page, 30 January 1996: 1–4.

Walker, R. B. (1991). 'Fauna and Flora Protection in New South Wales, 1866–1948', *Journal of Australian Studies*, 28 March: 17–28.

Wapner, P. (1995). 'In Defense of Banner Hangers: The Dark Green Politics of Greenpeace' in B. Taylor (ed.), *Ecological Resistance Movements: The Global Emergence of Radical and Popular Environmentalism*. New York: State University of New York Press.

Ward, M. (1997). 'Reasons for Decision of Magistrate Ward'. Delivered 18 February 1997 in the Magistrates Court at Canberra in the Australian Capital Territory: 1–18.

Webster, J. (1994). *Animal Welfare: A Cool Eye Towards Eden*. Oxford: Blackwell Science Ltd.

Weick, K. (1984). 'Small Wins: Redefining the Scale of Social Problems', *American Psychologist* 39 (1): 40–9.

Weinbren, D. (1994). 'Against All Cruelty: The Humanitarian League, 1891–1919', *History Workshop Journal* 38: 85–105.

Wiebers, D. O., Leaning, J., and White, R. D. (1994). 'Animal Protection and Medical Science', *The Lancet* 343 (9) (April): 902–4.

Wildlife Guardian. (1996). 'Hunters on the Run', *Journal of the League Against Cruel Sports* Summer, Issue 34.

Wiltfang, G., and McAdam, D. (1991). 'The Costs and Risks of Social Activism: A Study of Sanctuary Movement Activism', *Social Forces* 69 (4)(June): 987–1010.

Winsten, S. (1951). *Salt and His Circle*. London: Hutchinson & Co.

Withington, A. F. (1991). *Toward a More Perfect Union: Virtue and the Formation of American Republics*. New York: Oxford University Press.

Wolfe, A. (1993). *The Human Difference: Animals, Computers and the Necessity of Social Science*. Berkerley: University of California Press.

Worster, D. (1977). *Nature's Economy: The Roots of Ecology*. San Francisco: Sierra Club Books.

WSPA (1993). *Libearty News*, Issue No. 5, July. Animal Rights' Handbook (1996). *The Animals' Agenda* (July/August): 66–75.

WSPA (1996). Personal correspondence, Carole de Fraga, World Society for the Protection of Animals (Melbourne).

WSPA (1997). Personal correspondence, Stephanie Roderick, World Society for the Protection of Animals (Boston).

Zald, M. N. (1997). 'The Trajectory of Social Movements in America' in S. M. Buechler and F. K. Clyke (eds.) *Social Movements: Perspectives and Issues*. Mountain View, California: Mayfield Publishing Company.

Zelizer, V. (1985). *Pricing the Priceless Child: The Changing Social Value of Children*. New York: Basic Books.

Index

About the Author

LYLE MUNRO teaches Sociology at the School of Humanities, Communications, and Social Sciences at Monash University Gippsland. He has published a number of articles on the animal rights movement.